NUREG-1609
Supplement 1

Standard Review Plan for Transportation Packages for MOX-Radioactive Material

Manuscript Completed: September 2005
Date Published: September 2005

Prepared by
R.S. Hafner, G.C. Mok, J. Hovingh,
C. K. Syn, E.W. Russell, S.C. Keaton,
J.L. Boles, D.K. Vogt, P. Prassinos

Lawrence Livermore National Laboratory
7000 East Avenue
Livermore, CA 94550-9234

J.A. Smith, NRC Project Manager

Spent Fuel Project Office
Office of Nuclear Material Safety and Safeguards
U.S. Nuclear Regulatory Commission
Washington, D.C. 20555-0001
NRC Job Code A0291

ABSTRACT

The NRC contracted with LLNL to compile this supplement to NUREG-1609 to incorporate additional information specific to mixed uranium-plutonium oxide (MOX) radioactive material (RAM). This supplement provides details on package review guidance resulting from significant differences between low-enriched uranium (LEU) oxide radioactive material (RAM) and that from MOX RAM. While comparisons of MOX RAM with LEU RAM include powder, fresh-fuel rods, and fresh-fuel assemblies, they do not include comparisons with high-enriched uranium (HEU) or plutonium as metal, oxide, or nitrate in like forms that are covered by NUREG-1609. The principal purpose of this supplement is to ensure the quality and uniformity of staff reviews of packagings intended for transport of MOX RAM. It is also the intent of this plan to make information about regulatory matters widely available, and improve communications between NRC, interested members of the public, and the nuclear industry, thereby increasing the understanding of the NRC staff review process. In particular, this supplemental guidance, together with NUREG-1609, assists potential applicants by indicating one or more acceptable means of demonstrating compliance with the applicable regulations.

CONTENTS

LIST OF FIGURES

LIST OF TABLES

ACRONYMS AND ABBREVIATIONS

Code of Federal Regulations	CFR
fuel grade	FG
hypothetical accident conditions	HAC
low-enriched uranium	LEU
mixed uranium-plutonium oxide	MOX
maximum normal operating pressure	MNOP
normal conditions of transport	NCT
Nuclear Regulatory Commission	NRC
power grade	PG
radioactive materials	RAM
Regulatory Guide	RG
Safety Analysis Report	SAR
spontaneous fission	SF
weapon grade	WG
worst-case difference	WCD

INTRODUCTION

The Standard Review Plan for Transportation Packages for Radioactive Material (NUREG 1609)[1-1] provides guidance for the U.S. Nuclear Regulatory Commission (NRC) safety reviews of packages used in the transport of radioactive materials (RAM) under Title 10 of the U.S. Code of Federal Regulations (CFR), Part 71 (10 CFR Part 71). It is not intended as an interpretation of NRC regulations. NUREG-1609 supplements NRC Regulatory Guide (RG) 7.9, "Standard Format and Content of Part 71 Applications for Approval of Packaging for Radioactive Material,"[1-2] for review of package applications. NUREG-1609 involves guidance for reviewing radioactive material packagings intended for transport of radioactive materials, including low-enriched uranium, and low-enriched uranium oxide (LEU) in any form.

Following long-standing, existing regulatory requirements, it should be noted that 1) the transport of large quantities of plutonium, in liquid form, is not authorized, and 2) that air transport is not authorized (except for packages that meet 10 CFR 71.88).

This current report is not a stand-alone document but is intended as a supplement to NUREG-1609. This supplement to NUREG-1609 is intended to provide details on package review guidance resulting from the significant differences between contents of low-enriched uranium (LEU) oxide radioactive material (RAM) and contents of mixed uranium-plutonium oxide (MOX) RAM. While comparisons of MOX RAM with LEU RAM include powder, fresh-fuel rods, and fresh-fuel assemblies, they do not include comparisons with high-enriched uranium (HEU) or plutonium as metal, oxide, or nitrate in the forms of powder, pellets, fresh-fuel rods, or fresh fuel assemblies that are covered by NUREG-1609. Nothing contained in this document may be construed as having the force and effect of NRC regulations (except where the regulations are cited), or as indicating that applications supported by safety analyses and prepared in accordance with RG 7.9 will necessarily be approved, or as relieving any person from the requirements of 10 CFR Parts 20, 30, 40, 60, 70, or 71 or any other pertinent regulations. The principal purpose of this supplement to NUREG-1609 is to ensure the quality and uniformity of staff reviews of packagings intended for transport of MOX RAM. It is also the intent of this plan to make information about regulatory matters widely available and improve communications between NRC, interested members of the public, and the nuclear industry, thereby increasing the understanding of the NRC staff review process. In particular, this supplemental guidance, together with NUREG-1609, assists potential applicants by indicating one or more acceptable means of demonstrating compliance with the applicable regulations.

This supplement to NUREG-1609 is organized in the same manner as NUREG-1609, and has the identical numbering of subsections as found in that document. In addition, appendices found in this supplement are labeled to allow this report to be completely merged with NUREG-1609 without needing to change any labeling. For example, NUREG-1609 has two appendices labeled A and B with Appendix A being composed of eight parts. This supplement has four appendices, with two labeled A-9 and A-10, and the other two labeled C and D. Appendix A-9 contains information on a packaging for MOX powder and/or pellets, and Appendix A-10 contains information on a packaging for unirradiated MOX fuel (also referred to as MOX-fresh fuel in this document). Appendix C contains information on differences between thermal and radiation properties of MOX RAM and LEU RAM. Appendix D contains information on benchmark considerations for MOX RAM.

The subsection numbering structure within each section in NUREG-1609 is the same. The fifth subsection is labeled Review Procedures, and lists different review approaches for any subsection. These different review approaches in each Review Procedures subsection in this supplement are consequences of significant differences between LEU-RAM or MOX-RAM packages that potentially affect the compliance corresponding to the section of the Safety Analysis Report (SAR) in question with NRC regulations. If no significant differences exist for a particular subsection, that particular subsection is omitted from this supplement to NUREG-1609.

There are three generic differences that can affect each major section and potentially cause significant differences in review procedures between packages containing LEU RAM and MOX RAM. First, LEU RAM requires Type A Fissile Materials packagings, whereas MOX RAM requires Type B Radioactive Materials packagings, as specified by 49 CFR 173.431, 10 CFR 71.0(a)(2), and 10 CFR 71.51. Second, the containment requirements for LEU and MOX RAM are defined in 10 CFR 71.51, 10 CFR 71.55 and 49 CFR 173.24(b)(1). These features will be covered in more detail in Section 4 on containment.

Another of the potentially significant differences between LEU RAM and MOX RAM results because MOX RAM can have larger heat generation rates, photon emission rates, and neutron emission rates due to decay (see Appendix C for a discussion of these features). Another potentially significant difference between LEU RAM and MOX RAM results because the plutonium in MOX is a significant radiological hazard, and this can affect the allowable leakage requirements imposed on a package. In addition, several other differences are also noted in some of the sections that warrant review attention.

The U.S. DOE Standard DOE-STD-3013-2000 (herein called the 3013 Standard)[1-3] will be used to specify typical grades of plutonium employed to make the MOX-fresh fuel discussed in the text and in Appendix C. The actual plutonium compositions found in practice may not match these compositions exactly, but these grades can be considered typical for the purposes of this supplement to NUREG-1609. The 3013 Standard gives weight percents for various isotopes in various grades of plutonium. They are reproduced in the following table as representative values for typical grades of plutonium used to fabricate MOX-fresh fuel.

Table 1. Typical Isotopic Mix in Weight Percent for Various Grades of Plutonium as Specified in the 3013 Standard

Isotope	Weapon Grade	Fuel Grade	Power Grade
^{238}Pu	0.05	0.1	1.0
^{239}Pu	93.50	86.1	62.0[a]
^{240}Pu	6.00	12.0	22.0
^{241}Pu	0.40	1.6	12.0
^{242}Pu	0.05	0.2	3.0

Note: ^{236}Pu and ^{241}Am could be present but are not included in the 3013 Standard
[a] 63% reduced to 62% so that sum is 100%

Comparable information can also be found on the website of the American Nuclear Society.[1-4, 1-5]

References

1-1. U.S. Nuclear Regulatory Commission, "Standard Review Plan for Transportation Packages for Radioactive Material," NUREG-1609, U.S. Government Printing Office, Washington, D.C., 1999.

1-2. U.S. Nuclear Regulatory Commission, "Standard Format and Content of Part 71 Applications for Approval of Packaging for Radioactive Material," Regulatory Guide 7.9, Rev. 1, 1986.

1-3. U.S. Department of Energy, "Stabilization, Packaging, and Storage of Plutonium-Bearing Materials," U.S. DOE Standard, DOE-STD-3013-2000, Washington D.C., September 2000.

1-4. American Nuclear Society, Position Statement #42, *Protection and Management of Plutonium*, August 1995.

1-5. American Nuclear Society, Position Statement #47-bi, *Disposition of Surplus Weapons Plutonium Using Mixed Oxide Fuel*, Background Report, November 2002.

1 GENERAL INFORMATION REVIEW

1.5 Review Procedures

The general information review of NUREG-1609 is normally applicable to the review of both MOX-RAM and LEU-RAM packages. In this section, no significant deviations exist in the review procedures and considerations for the two packages. This section considers each of the subsections of Section 1.5 (Review Procedures) of NUREG-1609, and highlights the special considerations or attention needed for MOX-RAM packages. In subsections where no significant differences were found, that particular subsection has been omitted from this section.

For all packages, the general information review is based in part on the descriptions and evaluations presented in the Structural Evaluation, Thermal Evaluation, Containment Evaluation, Shielding Evaluation, Criticality Evaluation, Operating Procedures, and Acceptance Tests and Maintenance Program sections of the SAR. Similarly, results of the general information review are considered in the review of the SAR sections on Structural Evaluation, Thermal Evaluation, Containment Evaluation, Shielding Evaluation, Criticality Evaluation, Operating Procedures, and Acceptance Tests and Maintenance Program.

1.5.1 Introduction

Except for the need to use Type B packages, there should be no significant differences in the general methods to be used for review of LEU-RAM or MOX-RAM packages. Note that essentially all packages shipping bulk unirradiated MOX powder and pellets will be designated as Category I Packages by RG 7.11.

1.5.2 Package Description

Except for the possible need to use Type B packages, there should be no significant differences in the general methods to be used for review of LEU-RAM or MOX-RAM packages.

2 STRUCTURAL REVIEW

2.5 Review Procedures

The structural review of NUREG-1609 is normally applicable to the review of both MOX-RAM and LEU-RAM packages. However, details of the structural review procedures and considerations for LEU-RAM and MOX-RAM packages can be quite different. The differences arise from the generic differences in gamma and neutron emission rates of the two radioactive materials (see Appendix C). The presence of plutonium makes MOX RAM more of a radiological hazard than is LEU. The magnitude of the difference depends on the grade of plutonium in the contents. (See Table 1.) These conditions lead to the requirement for a Type B Fissile Materials package for containment of the MOX-RAM contents, while LEU packages are usually designated a Type A Fissile Materials package. This section considers each of the subsections of Section 2.5 (Review Procedures) of NUREG-1609, and highlights the special considerations or attention needed for the MOX-RAM packages. In subsections where no significant differences were found, that particular subsection has been omitted from this section.

For all packages, the structural review is based in part on the descriptions and evaluations presented in the General Information and the Thermal Evaluation sections of the SAR. Similarly, results of the structural review are considered in the review of the SAR sections on Thermal Evaluation, Containment Evaluation, Shielding Evaluation, Criticality Evaluation, Operating Procedures, and Acceptance Tests and Maintenance Program.

2.5.1 Description of Structural Design

The MOX-RAM package usually has a containment system while almost all LEU-RAM packages do not. In addition, there may be more neutron absorbers for criticality control and more shielding materials in the MOX-RAM package than in the LEU package. The structural reviewer should identify and understand the precise structural design and expected structural performance of all the packaging components that serve the containment, shielding, and criticality control functions.

2.5.1.2 Identification of Codes and Standards for Package Design

There should be no significant differences in the general methods to be used for review of LEU-RAM or MOX-RAM packages. However, the design code sections for the LEU-RAM package may not be appropriate for the MOX-RAM package, and the code sections for MOX-RAM packages can be over-restrictive for LEU-RAM packages.

2.5.2 Materials

2.5.2.1 Material Properties and Specifications

See also the additional information presented in Section X.5.2.4 of ISG-15.

2.5.2.2 Prevention of Chemical, Galvanic, or Other Reactions

Depending on the packaging materials, there can be significantly greater chemical reactions in a MOX-RAM package than in an LEU-RAM package. The structural reviewer should be alert in identifying undesirable conditions. Powder contents with high moisture are particularly susceptible to gas generation due to radiolysis, whereas LEU powder contents are not.

See also the additional information presented in Section X.5.3.1 of ISG-15.

2.5.2.3 Effects of Radiation on Materials

Depending on the packaging materials, there can be significantly greater radiation effects in a MOX-RAM package than in an LEU-RAM package. The structural reviewer should be alert in identifying undesirable conditions. Powder contents with high moisture are particularly susceptible to gas generation due to radiolysis, whereas LEU powder contents are not.

2.5.3 Fabrication and Examination

There should be no significant differences in the general methods to be used for review of LEU-RAM or MOX-RAM packages. However, the MOX-RAM and LEU-RAM packages may not need to use the same design code sections. The design code sections for the LEU-RAM package may not be appropriate for the MOX-RAM package, and the code sections for MOX-RAM packages can be over-restrictive for LEU-RAM packages.

2.5.4 Lifting and Tie-Down Standards for All Packages

2.5.4.1 Lifting Devices

There should be no significant differences in the general methods to be used for review of LEU-RAM or MOX-RAM packages. However, if the MOX-RAM package has a containment system in addition to fuel cladding, the failure of the lifting devices must not impair the ability of the containment system to meet its containment functions during normal conditions of transport (NCT) or hypothetical accident conditions (HAC).

2.5.4.2 Tie-Down Devices

There should be no significant differences in the general methods to be used for review of LEU-RAM or MOX-RAM packages. However, if the MOX-RAM package has a containment system in addition to fuel cladding, the failure of the tie-down devices must not impair the ability of the containment system to meet its containment functions during NCT or HAC.

2.5.6 Structural Evaluation under Normal Conditions of Transport

The only significant difference in the structural review of the LEU RAM and MOX RAM under NCT is in the containment acceptance criteria. Being a Type B package, the MOX-RAM package must meet the quantitative containment criterion specified in 10 CFR Part 71 for NCT. On the other hand, the Type-A LEU-RAM package does not need to meet any quantitative containment criteria.

2.5.7 Structural Evaluation under Hypothetical Accident Conditions

Similar to the structural review under NCT, the only significant difference in the structural review of the LEU RAM and MOX RAM under the HAC is in the containment acceptance criteria. A Type-A LEU-RAM package does not necessarily need to meet a quantitative release limit, whereas a Type-B MOX-RAM package does need to meet a quantitative release limit. Thus, the structural evaluation needs to demonstrate the structural integrity of the containment system under HAC for the MOX-RAM package but not for the LEU-RAM package. Note that the packaging of a LEU-RAM package, as well as a MOX-RAM package, must be robust enough to prevent criticality during an HAC.

References

2-1. U.S. NRC, Spent Fuel Project Office, Interim Staff Guidance -15 (ISG-15), "Materials Evaluation"

3 THERMAL REVIEW

3.5 Review Procedures

The thermal review of NUREG-1609 is normally applicable to the review of both MOX-RAM and LEU-RAM packages. In this section, there may be some differences in emphasis in the thermal review procedures that arise from generic differences between LEU-RAM and MOX-RAM packagings and contents. Plutonium has a higher specific activity of energetic, and short-ranged decay particles (~5 MeV alphas), than does LEU RAM. This results in higher specific content decay heat rates in the MOX-RAM packages than in other LEU-RAM packages (see Appendix C). Also MOX-fresh-fuel rods and assemblies may need special attention in some of the subsections. This section considers each of the subsections of Section 3.5 (Review Procedures) of NUREG-1609, and highlights the special considerations or attention needed for MOX-RAM packages. In subsections where no significant differences were found, that particular subsection has been omitted from this section.

For all packages, the thermal review is based in part on the descriptions and evaluations presented in the General Information and the Structural Evaluation sections of the SAR. Similarly, results of the thermal review are considered in the review of the SAR sections on Structural Evaluation, Containment Evaluation, Shielding Evaluation, Criticality Evaluation, Operating Procedures, and Acceptance Tests and Maintenance Program.

3.5.1 Description of Thermal Design

3.5.1.2 Content Decay Heat

There should be no significant differences in the general methods to be used for review of LEU-RAM or MOX-RAM packages. The decay heat generation rate for MOX RAM will be larger by up to four or five orders of magnitude (see Appendix C), than the usually negligible decay heat generation rate for LEU RAM.

3.5.1.3 Summary Tables of Temperatures

For MOX-fresh-fuel rods, the summary tables of the temperatures of package components including, but not limited to, the fuel/cladding, basket, impact limiters, containment vessel, seals, shielding, and neutron absorbers shall be consistent with the temperatures presented in the General Information and Structural Evaluation sections of the SAR for the NCT and HAC.

3.5.3 General Considerations for Thermal Evaluations

3.5.3.2 Evaluation by Test

For MOX-fresh-fuel rods and assemblies, temperature-sensing devices shall be placed on the fuel basket and fuel rods.

3.5.4 Thermal Evaluation under Normal Conditions of Transport

3.5.4.2 Maximum Normal Operating Pressure

For MOX-fresh-fuel rods and assemblies, the thermal evaluation shall determine the maximum normal operating pressure (MNOP) when the package has been subjected to the heat condition for one year. The reviewer should ensure that the evaluation has considered all possible sources of gases such as those present in the package at closure, and/or fill gas released from the MOX-fresh-fuel rods.
The evaluation of MOX powder and pellets on the MNOP should be similar to that of PuO_2 powder and pellets. For powders, however, it should be noted that there is the possibility that hydrogen, and/or other gases, may be produced from the thermal- or radiation-induced decomposition of the moisture associated with impure

plutonium-containing oxide powders. Given that the ratio of plutonium oxide powder to uranium oxide powder with respect to the total amount of MOX powder is expected to be small, any additional contributions from such gases should also expected to be small.

By the time the MOX powders are converted to fuel pellets, the processing temperatures should have removed all of the impurities from the plutonium oxide. From this point on (i.e., from MOX pellets, to MOX fuel rods, to full fuel assemblies), the evaluations of MOX pellets and LEU pellets should be virtually identical.

3.5.4.3 Maximum Thermal Stresses

For MOX-fresh-fuel rods and assemblies, the thermal stresses from temperature gradients through the components such as the containment vessel and fuel/clad shall be determined. These thermal stresses due to the temperature gradients will, in general, be small because the temperature gradients through the metal are small.

For MOX powders and fuel pellets, the processing temperatures should have removed all of the impurities from the plutonium oxide. From this point on (i.e., from MOX pellets, to MOX fuel rods, to full fuel assemblies), the evaluations of MOX pellets and LEU pellets should be virtually identical.

3.5.5 Thermal Evaluation under Hypothetical Accident Conditions

3.5.5.1 Initial Conditions

For MOX-fresh-fuel rods and assemblies, the internal heat load of the MOX-fresh-fuel contents shall be at its maximum allowable power, unless a lower power, consistent with the temperature and pressure, is more unfavorable.

For MOX powders and fuel pellets, the internal heat load of the contents shall be at its maximum allowable power, unless a lower power, consistent with the temperature and pressure, is more unfavorable.

3.5.5.2 Fire Test Conditions

For MOX-fresh-fuel rods and assemblies, the internal heat load of the MOX-fresh-fuel contents shall be at its maximum allowable power, unless a lower power, consistent with the temperature and pressure, is more unfavorable.

For MOX powders and fuel pellets, the internal heat load of the contents shall be at its maximum allowable power, unless a lower power, consistent with the temperature and pressure, is more unfavorable.

3.5.5.3 Maximum Temperatures and Pressures

For MOX-fresh-fuel rods and assemblies, the possible increases in gas inventory (e.g., from an unlikely failure of a fuel rod) shall be considered in the pressure determination.

For MOX powders and fuel pellets, the processing temperatures should have removed all of the impurities from the plutonium oxide. The only additional increase in pressure should be result of any helium released from the contents, as a result of the increased temperature. But, because any increase in temperature as a result of the thermal testing should be small when compared to the processing temperatures, any increase in pressure should be small.

3.5.6 Appendix

3.5.6.3 Applicable Supporting Documents or Specifications

For MOX-fresh-fuel rods and assemblies, the applicable sections from reference documents shall be included. These documents may include the test plans used for the thermal tests, the thermal specifications of O-rings, fuel clad, and other components, and the documentation of the thermal properties of non-ASME-approved materials used in the package.

Similar documentation should also be included for MOX powders and pellets.

4 CONTAINMENT REVIEW

4.5 Review Procedures

The containment review of NUREG-1609 is normally applicable to MOX-RAM and LEU-RAM packages. There are two major differences between the review approaches for LEU-RAM and MOX-RAM packagings: Unirradiated LEU packagings have traditionally come under the heading of Type A Fissile Materials packagings. MOX-RAM packagings, due to the intentional incorporation of the plutonium, can only be considered Type B Radioactive Materials packagings and requires at least one containment system, as defined in 10 CFR Part 71.

These differences will be discussed in more detail below. This section considers each of the subsections of Section 4.5 (Review Procedures) of NUREG-1609, and highlights the special considerations or attention needed for MOX-RAM packages. In subsections where no significant differences were found, that particular subsection has been omitted from this section.

For all packages, the containment review is based in part on the descriptions and evaluations presented in the General Information, Structural Evaluation, and Thermal Evaluation sections of the SAR. Similarly, results of the containment review are considered in the review of the SAR sections on Operating Procedures, and Acceptance Tests and Maintenance Program.

4.5.1 Description of the Containment System

4.5.1.1 Containment Boundary

The containment boundary information in section 4.5.1.1 of NUREG-1609 will have to be considered for each containment boundary for MOX packages. (See also the additional information provided in Section X.5.2.9 (Seals) of ISG-15.)

4.5.1.2 Special Requirements for Plutonium

The text of §71.63 of 10 CFR Part 71 states:

> "Shipments containing plutonium must be made with the contents in solid form, if the contents contain greater than 0.74 TBq (20 Ci) of plutonium."

4.5.2 General Considerations

4.5.2.1 Type A Fissile Packages

Because MOX-RAM packagings can only be considered Type B Radioactive Materials packagings, the containment-related information provided in Section 4.5.2.1 of NUREG-1609 can only be applied in the context of Type B Radioactive Materials packagings.

4.5.2.2 Type B Packages

Section 4.5.2.2 of NUREG-1609 specifies that Type B packagings must satisfy the quantified release rates in §71.51 of 10 CFR Part 71. An acceptable method for satisfying these requirements is provided in ANSI N14.5.[4-1] Additional information for determining containment criteria is provided in NUREG/CR-6487.[4-2]

4.5.3 Containment under Normal Conditions of Transport (Type B Packages)

4.5.3.1 Containment Design Criteria

Following the methodology suggested in NUREG/CR-6487, the containment design criteria for packagings designed to transport MOX-RAM powders and/or MOX-RAM pellets will typically be on the order of 10^{-6} reference cm^3/sec for NCT. The containment design criteria will have to be applied to each containment system independently. A full justification for the containment design criteria will have to be provided by the applicant in their calculations for all containment systems.

Although a leakage rate criterion on the order of 10^{-6} reference cm^3/sec may appear to be somewhat restrictive, leakage rates of this order of magnitude are actually relatively easy to meet in practice for packagings of this type. In many cases, the applicant may find it easier to default to the leaktight criterion specified in ANSI N14.5, i.e., $\leq 1 \times 10^{-7}$ reference cm^3/sec. Using this criterion also eliminates the applicant's need to fully justify containment design criteria calculations.

For packagings designed to transport MOX-fresh-fuel rods and/or MOX-fresh-fuel assemblies, which would not normally be considered to be dispersible radioactive material, the containment design criteria will be dependent on how the applicant defines the releasable source term in their calculations. Thus, although a single containment system will be specified for this type of packaging, the allowable leakage rate test criteria could be relatively high, i.e., $>10^{-3}$ reference cm^3/sec. For criteria as high as these, an applicant will have to fully justify their containment design criteria calculations. As an alternative, the staff has determined that an NCT containment design criterion of $\leq 1 \times 10^{-4}$ reference cm^3/sec is conservative for packagings of this type. If the applicant adopts this conservative criterion, the applicant is not required to provide any containment calculations to justify their position for NCT, beyond showing that the fuel remains largely intact such that a lower reference rate is not required.

4.5.4 Containment under Hypothetical Accident Conditions (Type B Packages)

4.5.4.1 Containment Design Criterion

The methodology presented in NUREG/CR-6487 suggests that the containment design criteria for packagings designed to transport MOX-RAM powders and/or MOX-RAM pellets could easily be higher for HAC than for NCT. In many cases, however, the applicant may find it useful to adopt the same leaktight containment design criteria for HAC defined above for NCT, i.e., $\leq 1 \times 10^{-7}$ reference cm^3/sec.

If the applicant adopts this conservative criterion, the applicant is not required to provide any containment calculations to justify their position for HAC. For those situations where the applicant adopts a higher criteria for HAC, the applicant must fully justify their containment design criteria calculations.

Because the methodology is, again, totally dependent on how the applicant defines the releasable source term in their calculations, the methodology presented in NUREG/CR-6487 suggests that the containment design criteria for packagings designed to transport MOX-fresh-fuel rods and/or MOX-fresh-fuel assemblies could easily fall into the 10^{-2} or higher reference cm^3/sec region for HAC. For these criteria, a full justification for the containment design criteria will have to be provided by the applicant in their calculations for HAC. As an alternative, the staff has determined that an HAC containment design criterion of $\leq 1 \times 10^{-4}$ reference cm^3/sec is conservative for packagings of this type. If the applicant adopts this conservative criterion, the applicant is not required to provide any containment calculations to justify their position for HAC.

References

4-1.   Institute for Nuclear Materials Management, "American National Standard for Radioactive Materials — Leakage Tests on Packages for Shipment," ANSI N14.5-1997, New York, NY, 1998.

4-2.   U.S. Nuclear Regulatory Commission, "Containment Analysis for Type B Packages Used to Transport Various Contents," NUREG/CR-6487, U.S. Government Printing Office, Washington, D.C., 1996.

4-3.   U.S. NRC, Spent Fuel Project Office, Interim Staff Guidance -15 (ISG-15), "Materials Evaluation"

5 SHIELDING REVIEW

5.5 Review Procedures

The shielding review of NUREG-1609 is normally applicable to the review of both MOX-RAM and LEU-RAM packages. In this section, however, a few significant deviations may exist in the review procedures and considerations for the two packages. This section considers each of the subsections of Section 5.5 (Review Procedures) of NUREG-1609 and highlights the special considerations or attention needed for MOX-RAM packages. In subsections where no significant differences were found, that particular subsection has been omitted from this section.

For all packages, the shielding review is based in part on the descriptions and evaluations presented in the General Information, Structural Evaluation, and Thermal Evaluation sections of the SAR. Similarly, results of the shielding review are considered in the review of the SAR sections on Operating Procedures and Acceptance Tests and Maintenance Program.

See also the additional information provided in Section X.5.2.6 of ISG-15.

5.5.2 Radiation Source

5.5.2.1 Gamma Source

Although the decay photon emission rate for MOX RAM can be larger than the decay photon emission rate for LEU RAM by one or more orders of magnitude (see Appendix C), there should be no significant differences in the general methods to be used for review of LEU-RAM or MOX-RAM packages. Appendix C includes information pertinent to gamma emission rates for various isotopes of some transuranic elements. In addition, Appendix C includes information on gamma emission rates from MOX RAM containing different grades of plutonium. Note, that ^{236}Pu, ^{238}Pu, ^{241}Pu and ^{241}Am should be included in the source term when present in the contents.

5.5.2.2 Neutron Source

One potentially significant difference in the review approach between LEU RAM and MOX RAM is that the neutron dose rate can be much larger than the gamma dose rate. This means that particular care is necessary to determine the appropriate neutron source strength. The contribution from (α,n) reactions can be large relative to spontaneous fission for MOX RAM. Depending on the methods used to calculate these source terms, the applicant might still determine the energy group structure independently for spontaneous fission and (α,n) reactions. However, it is generally necessary to include contributions from both spontaneous fission and (α,n) reactions. Also neutron multiplication effects can be important sources of additional neutrons and should be included in the shielding analysis (most modern radiation transport codes inherently produce multiplication neutrons). Appendix C includes information pertinent to neutron emission rates for various isotopes of some transuranic elements. In addition, Appendix C includes information on neutron emission rates from MOX RAM containing different grades of plutonium. Note, that ^{238}Pu, ^{239}Pu, ^{240}Pu, ^{242}Pu and ^{241}Am should be included in the source term when present in the contents.

5.5.4 Shielding Evaluation

Other than including a neutron source term in calculations along with a photon source term, there should be no significant differences in the general methods to be used for review of LEU-RAM or MOX-RAM packages.

6 CRITICALITY REVIEW

6.5 Review Procedures

The criticality review of NUREG-1609 is normally applicable to the review of both MOX-RAM and LEU-RAM packages. In this section, however, a few significant deviations may exist in the review procedures and considerations for the two packages. This section considers each of the subsections of Section 6.5 (Review Procedures) of NUREG-1609 and highlights the special considerations or attention needed for MOX-RAM packages. In subsections where no significant differences were found, that particular subsection has been omitted from this section.

For all packages, the criticality review is based in part on the descriptions and evaluations presented in the General Information, Structural Evaluation, and Thermal Evaluation sections of the SAR. Similarly, results of the criticality review are considered in the review of the SAR sections on Operating Procedures and Acceptance Tests and Maintenance Program.

See also the additional information provided in Section X.5.2.7 of ISG-15.

6.5.7 Benchmark Evaluations

6.5.7.1 Applicability of Benchmark Experiments

There are considerably fewer criticality benchmark experiments using MOX RAM than LEU RAM. Therefore, differences between the package and benchmarks may be more substantial for MOX RAM than for LEU RAM, so it might be more difficult to properly consider them. Appendix D discusses the availability of MOX-RAM benchmarks and their important characteristics from a criticality perspective. Appendix D also discusses how a reviewer might choose a set of appropriate MOX-RAM benchmarks.

6.5.7.2 Bias Determination

Because of the lack of criticality benchmark experiments using MOX RAM, assigning a bias value for benchmarks may be more difficult than for LEU RAM. Appendix D discusses MOX-RAM benchmarks and how a reviewer might determine a conservative bias value from comparisons between benchmark experiments and criticality calculations of the multiplication coefficient for those experiments. Appendix D also discusses how a reviewer might determine a conservative bias value for situations when the number of MOX-RAM benchmarks is less than desirable.

References

6-1. U.S. NRC, Spent Fuel Project Office, Interim Staff Guidance -15 (ISG-15), "Materials Evaluation"

7 OPERATING PROCEDURES REVIEW

7.5 Review Procedures

The operating procedures review of NUREG-1609 is normally applicable to the review of both MOX-RAM and LEU-RAM packages. In this subsection, however, two significant deviations exist in the review procedures and considerations for the two packages. This section considers each of the subsections of Section 7.5 (Review Procedures) of NUREG-1609, and highlights the special considerations or attention needed for MOX-RAM packages. In subsections where no significant differences were found, that particular subsection has been omitted from this section.

For all packages, the operating procedures review is based in part on the descriptions and evaluations presented in the General Information, Structural Evaluation, Thermal Evaluation, Containment Evaluation, Shielding Evaluation, Criticality Evaluation, and Acceptance Tests and Maintenance Program sections of the SAR.

7.5.1 Package Loading

7.5.1.2 Loading of Contents

There are two major differences between the review approaches for LEU-RAM and MOX-RAM packagings.

1. Unirradiated LEU packagings have traditionally come under the heading of Type A Fissile Materials packagings. MOX RAM packagings, due to the intentional incorporation of the plutonium, can only be considered Type B Radioactive Materials packagings.

2. Because they are Type A packagings, there are no leakage test requirements that are normally associated with LEU-RAM packagings. MOX-RAM packagings, on the other hand, due to the intentional incorporation of plutonium, can only be considered Type B radioactive materials packagings. These packagings must meet the leakage test requirements specified in 10 CFR 71.51. (See also Section 4, above.)

For purposes of the Operating Procedures review, the primary leakage test requirement that should be noted is the Preshipment Leakage Test. For MOX-RAM packagings, the Preshipment Leakage Test requirement will usually be 10^{-3} reference cm^3/sec, for all containment systems, prior to every shipment. The Preshipment Leakage Test rates should be determined in accordance with ANSI N14.5.

7.5.2 Package Unloading

7.5.2.2 Removal of Contents

There should be no significant differences in the general methods to be used for review of LEU-RAM or MOX-RAM packages.

References

7-1. Institute for Nuclear Materials Management, "American National Standard for Radioactive Materials — Leakage Tests on Packages for Shipment," ANSI N14.5-1997, New York, NY, 1998.

8 ACCEPTANCE TESTS AND MAINTENANCE PROGRAM REVIEW

8.5 Review Procedures

8.5.1 Acceptance Tests

The acceptance tests review of NUREG-1609 is normally applicable to the review of both MOX-RAM and LEU-RAM packages. In this subsection, however, significant deviations exist in the review procedures and considerations for the two packages. This section considers each of the subsections of Section 8.5.1 (Review Procedures) of NUREG-1609, and highlights the special considerations or attention needed for MOX-RAM packages. In subsections where no significant differences were found, that particular subsection has been omitted from this section.

For all packages, the acceptance tests review is based in part on the descriptions and evaluations presented in the General Information, Structural Evaluation, Thermal Evaluation, Containment Evaluation, Shielding Evaluation, Criticality Evaluation, and Operating Procedures sections of the SAR.

The guidance in NUREG-1609 is applicable to either LEU-RAM or MOX-RAM packagings. However, unirradiated LEU-RAM packagings have traditionally come under the heading of Type A Fissile Materials packagings, while MOX-RAM packagings come under the heading of Type B Radioactive Materials packagings. Although the information presented in Sections 8.5.1.1, 8.5.1.2, 8.5.1.3, 8.5.1.5, 8.5.1.6, and 8.5.1.7 of NUREG-1609 is applicable for either LEU-RAM or MOX-RAM packagings, it only applies in the context of Type A packaging for LEU-RAM or Type B packaging for MOX-RAM.

8.5.1.4 Leakage Tests

There are two major differences between the review approaches for LEU-RAM and MOX-RAM packagings.

1. Unirradiated LEU RAM packagings have traditionally come under the heading of Type A Fissile Materials packagings. MOX RAM packagings, due to the intentional incorporation of the plutonium, can only be considered Type B Radioactive Materials packagings.

2. Because they are Type A packagings, there are no leakage test requirements normally associated with LEU-RAM packagings. MOX-RAM packagings, on the other hand, must meet the leakage test requirements specified in 10 CFR 71.51.

As noted in Section 4, the leakage test requirements for packages used to transport MOX-fresh-fuel rods and MOX-fresh-fuel assemblies may be substantially different from the requirements for packages used for the transport of MOX powders and pellets — they can easily be different by several orders of magnitude. For purposes of the Acceptance Tests review, it is important to verify that the packaging containment system(s) is subjected to the Fabrication Leakage Test requirements specified in ANSI N14.5. The acceptable leakage test criteria should be consistent with those identified in the Containment Evaluation section (i.e., Chapter 4) of the SAR.

8.5.2 Maintenance Program

The maintenance program review of NUREG-1609 is normally applicable to the review of both MOX-RAM and LEU-RAM packages. However, in this subsection, a few significant deviations exist in the review procedures and considerations for the two packages. This section considers each of the subsections of Section 8.5.2 (Review Procedures) of NUREG-1609 and highlights the special considerations or attention needed for MOX-RAM packages. In subsections where no significant differences were found, that particular subsection has been omitted from this section.

For all packages, the maintenance program review is based in part on the descriptions and evaluations presented in the General Information, Structural Evaluation, Thermal Evaluation, Containment Evaluation, Shielding Evaluation, Criticality Evaluation and Operating Procedures sections of the SAR.

The guidance in NUREG-1609 is applicable to either LEU-RAM or MOX-RAM packagings. However, unirradiated LEU-RAM packagings have traditionally come under the heading of Type A Fissile Materials packagings, while MOX-RAM packagings come under the heading of Type B Radioactive Materials packagings. Although the information presented in Sections 8.5.2.1, 8.5.2.3, 8.5.2.4, and 8.5.2.5 of NUREG-1609 is applicable for either LEU-RAM or MOX-RAM packagings, it only applies in the context of Type A packaging for LEU-RAM or Type B packaging for MOX-RAM.

8.5.2.2 Leakage Tests

There are two major differences between the review approaches for LEU-RAM and MOX-RAM packagings.

1. Unirradiated LEU packagings have traditionally come under the heading of Type A Fissile Materials packagings. MOX RAM packagings, due to the intentional incorporation of the plutonium, can only be considered Type B Radioactive Materials packagings.

   Seal replacement procedures for Type A Fissile Materials packagings should follow typical Type A packaging procedures, with no additional leakage testing requirements normally specified. Seal replacement procedures for Type B Radioactive Materials packagings, on the other hand, should follow typical Type B packaging procedures, with additional leakage tests specified for all containment boundary seals. For Type B packagings, the requirements specified should be in agreement with the requirements specified in Section 4, and all leakage test requirements should be agreement with the requirements of ANSI N14.5.[8-1]

2. Because they are Type A packagings, there are no leakage test requirements normally associated with LEU-RAM packagings. MOX-RAM packagings, on the other hand, must meet the leakage test requirements specified in 10 CFR 71.51.

   Seal replacement procedures for Type A Fissile Materials packagings should follow typical Type A packaging procedures, with no additional leakage testing requirements normally specified. Seal replacement procedures for Type B Radioactive Materials packagings, on the other hand, should follow typical Type B packaging procedures, with additional leakage tests specified for all containment boundary seals. For Type B packagings, the requirements specified should be in agreement with the requirements specified in Section 4, and all leakage test requirements should be agreement with the requirements of ANSI N14.5.[8-1]

As was noted in Section 4, the leakage test requirements for packages used for the transport of MOX-fresh-fuel rods and MOX-fresh-fuel assemblies may be substantially different from the requirements for packages used for the transport of MOX powders and pellets — they can easily be different by several orders of magnitude. For purposes of the Maintenance Program review, it is important to verify that the packaging containment system(s) is subjected to the Maintenance Leakage Test and/or the Periodic Leakage Test requirements specified in

ANSI N14.5. The acceptable leakage test criteria should be consistent with those identified in the Containment Evaluation section (i.e., Chapter 4) of the SAR.

References

8-1.   Institute for Nuclear Materials Management, "American National Standard for Radioactive Materials — Leakage Tests on Packages for Shipment," ANSI N14.5-1997, New York, NY, 1998.

APPENDICES

APPENDIX A9: MOX POWDER AND PELLET PACKAGES

A9.1 Package Type

A9.1.1 Purpose of Package

The purpose of this type of package is to transport Type B quantities of MOX material (other than by air).

A9.1.2 Description of a Typical Package

A typical packaging consists of a containment vessels and an outer container that serves to confine the package internals. The outer container (confinement system) is a steel drum with a removable head and weather-tight gasket. The head usually is a bolted or clamped lid with a tamperproof seal. Vent holes near the top of the drum, which provide pressure relief from combustion gases or off-gassing from insulating materials under HAC, are capped or taped during NCT to prevent water inleakage.

The inner containment vessel is a steel container, typically a stainless steel cylinder, with a maximum inner diameter of 0.127 m (5 in.), closed by a welded bottom cap and a welded top flange with a bolted lid. The lid, which is generally sealed by two O-rings, contains a leak-test port and sometimes a separate fill port for leak testing.

A secondary containment vessel or product container may be used and may be designed similar to the primary containment vessel; and can include welded and bolted bottom cap and top flange, respectively; dual O-ring seals; a leak test port; and sometimes a separate fill port for leakage testing. (See, for example, Figure A9-1.)

The contents are MOX powder or pellets. The MOX powder or pellets are generally placed in metal cans prior to loading into the containment vessel. Solid spacers are often used to maintain the position of the contents.

A sketch of a typical package with an optional containment vessel is shown in Figure A9-1.

A9.2 Package Safety

A9.2.1 Safety Functions

The principal functions of the package are to provide containment, shielding, and criticality control. Package design features that accomplish the containment and criticality functions might also provide adequate shielding to satisfy the requirements for nonexclusive-use shipment. Additional shielding may be required if significant quantities of certain isotopes, e.g., ^{236}Pu, ^{238}Pu, ^{241}Pu or ^{241}Am (from the decay of ^{241}Pu) are present in the MOX material.

A9.2.2 Safety Features

- The steel drum and thermal insulating/impact absorbing material protect the containment vessel and contents under HAC and maintain a minimum spacing between packagings for criticality control.

- The inner vessel provides containment of the radioactive material.

- The diameter and volume of the inner containment vessel, together with limits on the fissile mass of the contents, ensure that a single package is subcritical, even with water inleakage.

- The containment vessel, thermal insulating/impact absorbing material, and steel drum maintain a minimum distance from the contents to the package surface and provide some attenuation to satisfy the shielding requirements.

A9.2.3 Typical Areas of Review for Package Drawings

- Containment vessel body

    - Materials specifications

    - Dimensions and tolerances, including maximum cavity dimensions

    - Fabrication codes or standards

    - Weld specifications, including codes or standards for nondestructive examination.

- Containment vessel closures

    - Lid material specifications, dimensions, and tolerances

    - Bolt specifications, including number, size, material, and torque

    - Seal material specifications and size

    - Seal groove dimensions

    - Leak-test ports

    - Applicable codes and standards.

- Spacers to position or displace fissile material

    - Material of construction

    - Dimensions and tolerances

    - Locations.

- Thermal insulating/impact absorbing material

    - Type, and specifications

    - Dimensions and tolerances

    - Density.

- Outer drum

    - Material specifications, including lid and closure device

    - Closure bolt specifications, including number, size, material, and torque

    - Dimensions

    - Applicable codes or standards.

- Neutron poisons

A9.2.4 Typical Areas of Safety Review

- The structural review confirms that packaging integrity is maintained under both NCT and HAC, particularly, the drop, crush, and puncture tests. The review also verifies that the drum lid remains securely in place and the drum body and closure have no unacceptable openings that would cause the safety performance of the package to not meet regulatory standards, especially during the fire test.

- The structural and thermal reviews evaluate the performance of the containment system under both NCT and HAC. Primary emphasis is on the structural integrity of the containment vessel and its closure, and on the thermal performance of the O-rings.

- The structural and thermal reviews address the condition of the package and the minimum spacing between different packages under HAC. Damage to the outer drum and charring of the thermal insulating/impact absorbing material may result in closer spacing than that of NCT.

- The thermal and containment reviews verify that the hydrogen concentration in any confined volume will not exceed 5% (by volume) during a period of one year. Shorter time periods have been approved based on detailed operating procedures to control and track the shipment of packages.

- The shielding review evaluates the ability of the package to satisfy the allowed radiation levels during both NCT and HAC.

- The criticality review addresses, in detail, both NCT and HAC. Key parameters for this review include the number of packages in the arrays, array configuration (pitch, orientation of packages, etc.), positioning of the containment vessel within the drum, moderation due to inleakage of water, the condition and quantity of spacing material, interspersed moderation between packages, preferential flooding of different regions within the package, packaging materials that provide moderation (e.g., plastics), and neutron poisons.

- The review of operating procedures confirms that the containment vessel has been properly closed and its closure bolts are properly tightened to the specified torqued values, and that an appropriate pre-shipment leak test is performed.

- The review of the acceptance tests and maintenance program verifies that appropriate fabrication and periodic verification leakage tests are performed.

- The review of the acceptance tests and maintenance program also verifies that neutron poisons, if any, are present, and are subject to the appropriate tests to verify concentration.

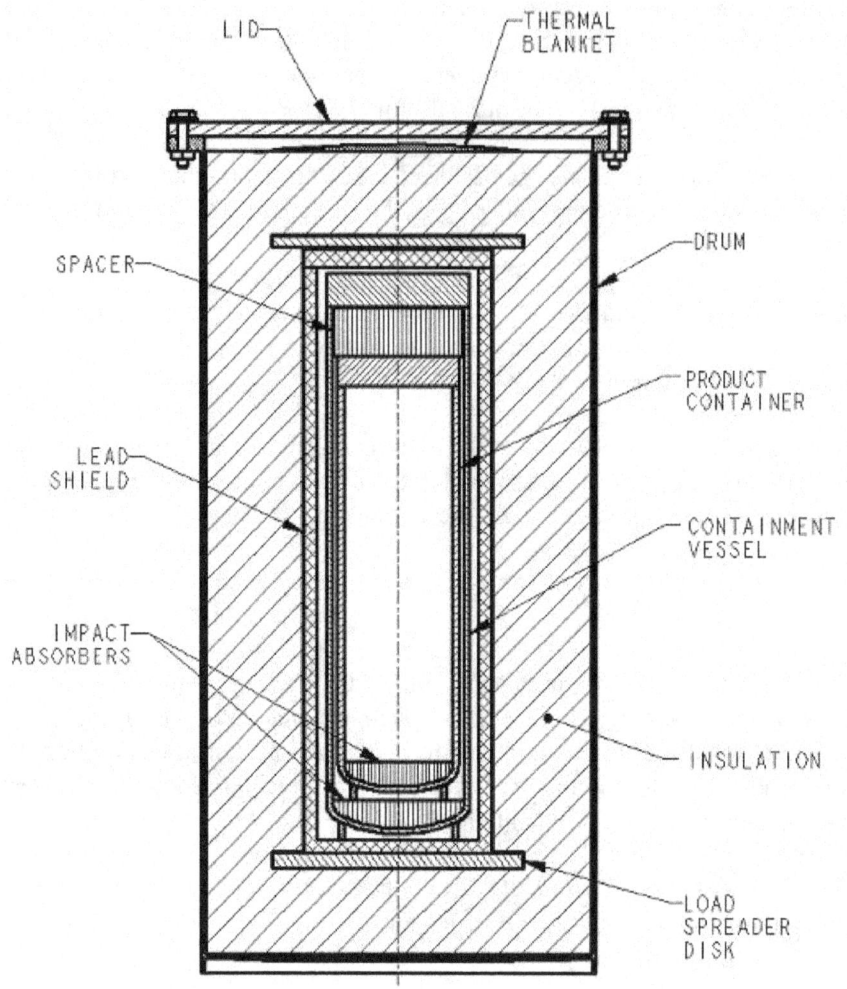

Figure A9-1. MOX Powder/Pellet Package

APPENDIX A10: UNIRRADIATED MOX FUEL PACKAGES

A10.1 PACKAGE TYPE

A10.1.1 Purpose of Package

The purpose of this type of package is to transport unirradiated MOX-fuel assemblies and individual MOX-fuel rods. These packages are also referred to as "MOX fresh-fuel packages."

This appendix addresses those packages in which the contents are Type B quantities of fissile MOX material. The fissile MOX material can be in an entire assembly, or as individual fuel rods.

A10.1.2 Description of a Typical Package

A typical packaging consists of a metal outer shell, closed with bolts and elastomeric seals, and an impact-limiter system. An internal steel strongback, shock-mounted to the outer shell, supports one or more fuel assemblies, which are fixed in position on the strongback by clamps, separator blocks, and end support plates. Depending on the type of fuel, neutron poisons are sometimes used to reduce reactivity. Material surrounding the contents could be employed to shield against neutrons and/or gammas. If the package is used to transport individual fuel rods, a separate inner container is often employed.

The contents of the package are unirradiated MOX in fuel assemblies or individual fuel rods. Because the majority of these packages are for commercial reactor fuel, the MOX is typically in the form of Zircaloy or stainless steel-clad plutonium-uranium dioxide pellets.

A sketch of the typical package described above is shown in Figure A10-1.

A10.1.3 Alternative Package Design

In an alternative design for a MOX fresh-fuel package, the fuel assemblies are fixed in position by two or three steel channels, mounted by angle irons or a similar bracing structure to a thin-walled inner metal container. This inner container is in turn surrounded by a honeycomb material and enclosed in a metal outer shell. Foam cushioning material can be used to cushion the fuel assemblies and may be used between the inner and outer container.

A10.2 PACKAGE SAFETY

A10.2.1 Safety Functions

The principal functions of the package are to provide containment, shielding, and criticality control. Package design features that accomplish the containment and criticality functions might also provide adequate shielding to satisfy the requirements for nonexclusive-use shipment. Additional shielding may be required if significant quantities of certain isotopes, e.g., ^{236}Pu, ^{238}Pu, ^{241}Pu or ^{241}Am (from the decay of ^{241}Pu) are present in the MOX material.

A10.2.2 Safety Features

- Impact limiters protect the outer shell and contents under HAC. They also provide thermal insulation for the O-ring seals of the outer shell.

- A strongback with end support plates, clamps, and separators, maintains the fuel assemblies in a fixed position relative to each other and to any neutron poisons.

- The metal outer shell of the packaging retains and protects the fuel assemblies, and may provide a minimum spacing between assemblies in an array of packages and provide some attenuation to satisfy the shielding requirements.

- Neutron poisons, if present, reduce reactivity, and can provide some neutron shielding.

- The metal outer shell also provides containment of the radioactive material.

A10.2.3 Typical Areas of Review for Package Drawings

- Outer shell (containment vessel body)

  - Material specifications

  - Dimensions and tolerances

  - Fabrication codes and standards

  - Weld specifications, including codes or standards for nondestructive examination.

- Outer shell closure (containment vessel closure)

  - Lid materials, dimensions, and tolerances

  - Bolt specifications, including number, size, and torque

  - Seal material, size, and compression specifications

  - Seal groove dimensions

  - Leak-test ports

  - Applicable codes and standards.

- Structural components (e.g., strongback, support plates, fuel clamps, and separators) that fix the position of fuel assemblies or relative position between fuel assemblies and poisons

  - Dimensions and material specifications

  - Methods of attachment

  - Applicable engineering codes or standards.

- Thermal insulating/impact absorbing and/or shielding material

  - Type and specifications

  - Dimensions and tolerances

  - Density.

- Neutron poisons
  - Dimensions and tolerances
  - Minimum poison content
  - Location and method of attachment
  - Material specifications
  - Applicable codes and standards.

- Moderating materials, including plastics, wood, and foam
  - Location
  - Material properties.

A10.2.4 Typical Areas of Safety Review

- The general information review identifies the fuel assembly designs authorized in the package, including:
  - Number of, and arrangement of fuel assemblies
  - Number, pitch, and position of fuel rods, guide tubes, and channels
  - Overall assembly dimensions, including active fuel length
  - Authorization or restrictions on missing fuel rods or partial-length rods
  - Maximum amount of fissile material
  - Pellet dimensions and tolerances
  - Minimum cladding thickness
  - Fuel-clad gap and fill gas
  - Type, location, and concentration of burnable poisons, and other types of poisons
  - Type, location, and quantity of plastics, such as polyethylene, within the fuel assemblies.

- The structural and thermal reviews evaluate the performance of the containment system under both NCT and HAC, particularly, the drop, crush (if needed), and puncture tests. Primary emphasis is on the structural integrity of the outer shell (containment vessel) and its closure, and on the thermal performance of the elastomeric seals. If the impact limiters provide thermal protection for the seals, the structural review also confirms the structural integrity of the impact limiters.

- The structural review addresses possible damage to the impact limiters, outer shell, strongback, fuel assembly, neutron poisons (if present), clamps, separators, and end support plates to ensure that the fuel assemblies and neutron poisons are maintained in a fixed position relative to each other under hypothetical accident conditions.

- The structural review also confirms the minimum spacing between fuel assemblies in different packages in an array under hypothetical accident conditions. Spacing can be affected by separation of the strongback from its shock mounts, failure of the shock mounts or fuel assembly clamps, and deformation of the outer shell of the package.

- The thermal review evaluates the effect of the fire on outer-shell O-ring seals, neutron poisons, plastic sheeting, thermal insulation material (if present), or other temperature-sensitive materials under hypothetical accident conditions.

- The structural and thermal reviews address the condition of the package and the minimum spacing between different packages under HAC. Damage to the outer shell and charring of any thermal insulating/impact absorbing material (if present) may result in closer spacing than that of NCT.

- The criticality review addresses both normal conditions of transport and hypothetical accident conditions. Key areas for this review include:

    - The number of packages in the array and the array configuration (pitch, orientation of packages, etc.). Because of movement of the strongback within the package and the location of poisons, the arrays might not be symmetrical.

    - Degree of moderation. Structural features, as well as packaging material such as plastic sheeting, are evaluated for the possibility of differential flooding within the package. Plastic sheeting on the fuel assemblies should be open at both ends to preclude differential flooding. Flooding between the fuel pellets and cladding is also considered. Variations in the allowable amount of lightweight packaging material and plastic shims inserted in the fuel assemblies can also affect criticality under normal conditions of transport.

- The shielding review evaluates the ability of the package to satisfy the allowed radiation levels during both NCT and HAC.

- The review of operating procedures ensures that instructions are provided so that proper clamps, separators, and poisons are selected for the type of fuel assemblies to be shipped and that these items are properly installed prior to shipment. The procedures should also address any other restrictions (e.g., limits on number of shims) considered in the package evaluation. The review also confirms that instructions are provided for the proper closure of the outer shell and for the proper completion of pre-shipment leak test.

- The review of the acceptance tests and maintenance program verifies that the neutron poisons, if present, are subject to appropriate tests to verify their concentration and uniformity. The review also verifies that appropriate fabrication and periodic verification leakage tests of the outer shell are performed.

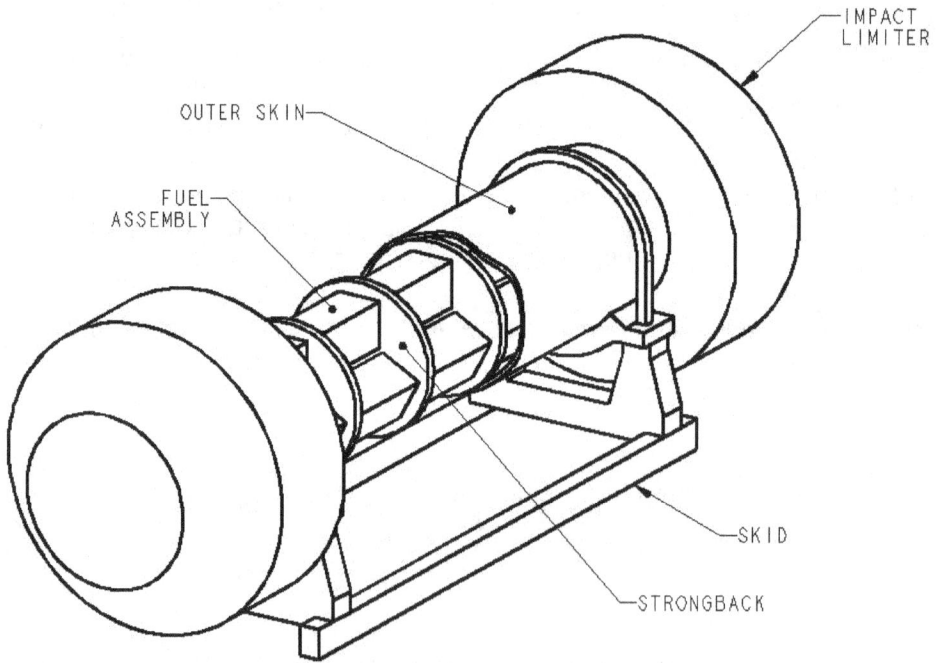

Figure A10-1. MOX Fresh Fuel Package

APPENDIX C: DIFFERENCES BETWEEN THERMAL AND RADIATION PROPERTIES OF MOX AND LEU RADIOACTIVE MATERIALS

The contents considered in this SRP are unirradiated MOX RAM, in the form of powder, pellets, fresh-fuel rods, or fresh-reactor fuel assemblies. Unirradiated MOX RAM will also be referred to in this appendix as MOX-fresh fuel. This appendix summarizes the relative degree of differences between the thermal and radiation properties of the various MOX-RAM contents relative to similar properties for analogous LEU-RAM contents. We will use the 3013 Standard[C-1] to specify typical grades of plutonium used to make the MOX fresh fuel discussed in this appendix. The actual plutonium compositions found in practice may not match these compositions exactly, but these grades can be considered typical for the purposes of this appendix.

The 3013 Standard gives weight percents for various plutonium isotopes in various grades of plutonium. They are reproduced in the following table as representative values for typical grades of plutonium that might be used to fabricate MOX-fresh fuel. Pure ^{239}Pu has been included to contrast the effect of the other plutonium isotopes. Note that in addition to the isotopes identified in Table C-1, plutonium will contain ^{236}Pu and ^{241}Am (from ^{241}Pu decay).

Table C-1.   Typical Isotopic Mix in Weight Percent for Various Grades of Plutonium as Specified in the 3013 Standard

Isotope	Pure ^{239}Pu	Weapon Grade	Fuel Grade	Power Grade
^{238}Pu	0	0.05	0.1	1.0
^{239}Pu	100	93.50	86.1	62.0[a]
^{240}Pu	0	6.00	12.0	22.0
^{241}Pu	0	0.40	1.6	12.0
^{242}Pu	0	0.05	0.2	3.0

[a] 63% reduced to 62% so that sum is 100%

Initially, it is expected that MOX-fresh fuel will be fabricated using Weapon Grade (WG) plutonium. A more mature MOX-fuel program might be expected to fabricate MOX-fresh fuel from previously irradiated WG-MOX fuel that may have a composition similar to Fuel Grade (FG) plutonium. Fabricating MOX-fresh fuel from Power Grade (PG) plutonium would require a much more mature MOX-plutonium program.

To compare MOX-fresh fuel with LEU-fresh fuel, we need to choose representative compositions for each fuel type. For a reference LEU-fresh fuel, we choose UO_2 with 4 wt% ^{235}U and 96 wt% ^{238}U. For the various grades of plutonium in MOX-fresh fuel, we choose UO_2-PuO_2 having 4 wt% ^{239}Pu with the remaining plutonium isotopes scaled as required by Table C-1, with the remainder being depleted uranium with 0.2 wt% ^{235}U and 99.8 wt% ^{238}U. The actual composition of MOX RAM found in practice will not match these compositions, but they are appropriate for comparing the effects of MOX RAM using various grades of plutonium. Table C-2 lists the weight percents for heavy metal isotopes used in this study.

Table C-2.   Weight Percents for Heavy Metal Isotopes Chosen for
Comparing MOX with LEU for Various Grades of Plutonium

Nuclide	No Plutonium[a]	Pure ^{239}Pu	Weapons Grade[b]	Fuel Grade[b]	Power Grade[b]
^{235}U	4.0000	0.1920	0.1914	0.1907	0.1871
^{238}U	96.0000	95.8080	95.5305	95.1653	93.3613
^{238}Pu	0.0000	0.0000	0.0021	0.0047	0.0645
^{239}Pu	0.0000	4.0000	4.0000	4.0000	4.0000
^{240}Pu	0.0000	0.0000	0.2567	0.5575	1.4194
^{241}Pu	0.0000	0.0000	0.0171	0.0743	0.7742
^{242}Pu	0.0000	0.0000	0.0021	0.0093	0.1935

[a] No plutonium means low-enriched uranium oxide with 4 wt% ^{235}U and 96 wt% ^{238}U. Note that fresh LEU fuel will normally contain traces of ^{232}U, ^{233}U, ^{234}U, and ^{236}U from recycle and natural uranium. The quantities of these isotopes normally present in fresh LEU are not significant for the comparisons in this appendix.
[b] The plutonium mixtures will also contain ^{236}Pu and ^{241}Am. These isotopes can have a significant effect on neutron and/or gamma generation rates.

The nuclide depletion/decay code ORIGEN-ARP[C-2] can be used to determine the heat generation rates for arbitrary compositions of plutonium with depleted uranium in MOX-fresh fuel. Table C-3a lists the ratio of heat generation rates for MOX-fresh fuel relative to LEU-fresh fuel using the composition weight percents for MOX-fresh fuel fabricated from the various plutonium grades from Table 2 in ORIGEN-ARP. These are the values predicted at the initial time of MOX fuel fabrication when the composition weight percents for the various plutonium isotopes are as given in Table C-2. After these nuclides begin to decay, the heat generation rate decreases with time, so the initial heat generation rate is also the maximum rate.

Table C-3a.   Ratio of Heat Generation Rate for MOX-Fresh Fuel Composed of
Various Grades of Plutonium Relative to LEU Fresh Fuel

Decay Time	No Plutonium	Pure ^{239}Pu	Weapons Grade	Fuel Grade	Power Grade
Initial	1	7,300	10,200	13,700	53,900
Maximum	1	7,300	10,200	13,700	53,900

The heat generation rate for any MOX-fresh fuel is about four orders of magnitude, or more, greater than that from LEU-fresh fuel. Using FG plutonium instead of WG plutonium causes the heat generation rate to increase by about another factor of 1.3. Using PG plutonium instead of FG plutonium causes the heat generation rate to increase by about another factor of 3.9. For reference, one metric ton of heavy metal of MOX fuel fabricated from WG plutonium will generate more than 100 watts of decay heat.

The heat is generated predominately by alpha decay of the heavy nuclides. The average alpha energy spectrum for the plutonium isotopes is greater than that for the uranium isotopes by about 25%. However, the primary reason heat generation is greater for plutonium is that its specific activity for alpha decay is four to five orders of magnitude larger that that for uranium. Some specific decay parameters for MOX-relevant are shown below, in Table C-3b.

Table C-3b.   Specific Decay Parameters for MOX-Relevant Nuclides

Radio-nuclide	Half-Life (Years)	Decay Energy (Mev/Event)	Decay Energy (Watt-yr/mole)	Specific Heat Generation Rate (Watts/kg)
^{233}U	1.60E+05	4.909	15,021	5.81E-01
^{235}U	7.10E+08	4.681	14,333	6.00E-05
^{238}U	4.50E+09	4.195	12,836	8.00E-06
^{238}Pu	8,78E+01	5.593	17,113	5.67E+02
^{239}Pu	2.41E+04	5.244	16,046	1.93E+00
^{240}Pu	6.54E+03	5.255	16,079	7.10E+00
^{241}Pu	1.44E+01	0.0205	62.7	1.25E+01
^{242}Pu	3.76E+05	4.983	15,246	1.16E-01
^{241}Am	4.32E+02	5.637	17,248	1.15E+02

The gamma emission code GAMGEN[C-3] can be used to determine the gamma emission rates for equal weights of various nuclides of uranium, plutonium, and americium. Shielding for LEU is not a significant problem as a function of decay time. Therefore, studying the gamma emission rate for each nuclide of interest relative to LEU gives a measure of how much more difficult shielding problems might be when that nuclide is present than for LEU. Table C-4a lists the gamma emission rates at twenty years decay time for equal weights of each nuclide, relative to the LEU gamma emission rate at twenty years decay time, for four energy ranges corresponding to different minimum gamma energies. Although gamma emission rates are not necessarily maximized at twenty years decay time, this decay time was chosen because it gives a better indication of the relation of the various nuclide emission rates relative to LEU with time. The maximum gamma energies for each nuclide are below 3.3 MeV, and sometimes significantly below. The reason the gamma emission ratios are listed for several different energy ranges is to provide some indication of the energy distribution for the gammas of each nuclide as the minimum gamma energy increases, since shielding becomes more difficult as gamma energy increases. This is facilitated by listing the average gamma energy for each nuclide for each energy range in the table. The fact that each nuclide has a different average gamma energy for a given energy range is because each has a unique gamma energy spectrum. When the average energy for a nuclide is close to the minimum energy for an energy range, this indicates that most gammas in that range have energies near to that of the minimum energy.

The nuclides ^{236}Pu and ^{232}U have very large emission ratios because of the relatively short half-lives and 2.614 MeV gammas emitted after chain decaying to ^{208}Tl. These gammas are extremely difficult to shield against and can usually be tolerated at amounts no greater than about 10^{-4} weight percent of heavy metal nuclides. The nuclides ^{236}U, ^{241}Am, ^{234}U, and ^{237}Np result from radioactive decay of ^{240}Pu, ^{241}Pu, ^{238}Pu, and ^{241}Am, respectively. The nuclide ^{233}U is usually present in trace quantities.

In Table C-4a for the minimum gamma energies corresponding to 0.041 and 0.183 MeV, most of the nuclides have emission ratios greater than 1.00. The nuclides ^{238}Pu, ^{240}Pu, ^{241}Pu, ^{242}Pu, ^{234}U, ^{236}U, and ^{241}Am have a majority of gammas in the energy range between roughly 0.04 and 0.12 MeV, because their average energies for the first energy range are close to the minimum energy of 0.041 MeV. However, except for ^{235}U, all nuclides have average energies greater than 0.28 MeV for the second energy range. Therefore, these nuclides have considerable gammas with energies that will require specific gamma shielding if present in sufficient quantities. This is reinforced by the emission ratios and average energies for the energy range with minimum gamma energy of 0.498 MeV, particularly for the plutonium isotopes. For the energy range with minimum gamma energy of 1.000 MeV, only ^{236}Pu (except for trace nuclides) has high emission rates of very high-energy gammas that may require substantial shielding if it is present in a significant quantity.

Table C-4a. Gamma Emission Rates Relative to the LEU Gamma Emission Rate and Average Gamma Energies for Equal Weights of Some Nuclides of Uranium, Plutonium, Neptunium and Americium at 20 Years Decay Time

Select Nuclides	Gamma Energies ≥0.041 MeV		Gamma Energies ≥0.183 MeV		Gamma Energies ≥0.498 MeV		Gamma Energies ≥1.000 MeV	
	Emission Relative to LEU	Average Energy (MeV)	Emission Relative to LEU	Average Energy (MeV)	Emission Relative to LEU	Average Energy (MeV)	Emission Relative to LEU	Average Energy (MeV)
^{236}Pu	2.73E+08	0.7929	4.46E+08	0.9927	2.90E+09	1.4300	2.26E+09	2.5212
^{238}Pu	5.33E+04	0.0624	1.37E+02	0.7497	1.31E+03	0.7906	7.71E+01	1.1753
^{239}Pu	2.89E+02	0.1173	7.08E+01	0.3883	8.39E+00	0.6968	1.06E-02	1.1750
^{240}Pu	9.05E+02	0.0600	1.66E+00	0.3941	5.97E+00	0.6831	4.05E-09	2.1790
^{241}Pu	5.71E+06	0.0544	4.74E+03	0.2805	2.64E+03	0.6771	8.49E-07	1.4577
^{242}Pu	1.32E+01	0.0613	3.94E-06	0.9783	3.75E-05	1.0389	3.75E-05	1.2507
^{232}U	2.73E+08	0.7930	4.46E+08	0.9927	2.91E+09	1.4300	2.26E+09	2.5212
^{233}U	2.62E+02	0.1801	1.88E+02	0.3549	8.32E+01	1.3620	1.30E+02	1.4607
^{234}U	7.58E+01	0.0789	2.21E-01	0.7588	1.39E+00	1.0282	1.05E+00	1.5553
^{235}U	1.75E+01	0.1901	2.24E+01	0.2402	7.27E-03	0.7422	1.49E-04	1.1750
^{236}U	4.76E-01	0.0723	1.52E-06	0.8876	1.05E-05	1.1865	5.49E-06	2.2074
^{238}U	3.12E-01	0.2289	1.09E-01	0.9783	1.04E+00	1.0389	1.04E+00	1.2507
LEU	1.00E+00	0.2017	1.00E+00	0.3177	1.00E+00	1.0388	1.00E+00	1.2507
^{237}Np	6.06E+03	0.2096	5.94E+03	0.3374	2.63E-04	1.3575	4.09E-04	1.4597
^{241}Am	9.09E+06	0.0543	1.85E+03	0.3905	4.21E+03	0.6771	4.23E-06	1.4586

ORIGEN-ARP also gives the gamma emission rates for arbitrary compositions of plutonium with depleted uranium in MOX-fresh fuel. Table C-4b lists the ratio of gamma emission rates for MOX-fresh fuel relative to LEU-fresh fuel using the composition weight percents for MOX-fresh fuel fabricated from the various plutonium grades from Table 2 in ORIGEN-ARP. Table C-4b lists rates for both initial time and maximum rates after some decay time. The decay time at maximum gamma emission rates depends on the plutonium grade in question. The gamma emission rates include only gammas with energies equal to or greater than 100 keV. The assumption is that gammas with energies less than 100 keV will be absorbed by the normal packaging materials required to transport MOX-fresh fuel contents, specifically the strong 59.5 keV gammas coming from any ^{241}Am produced through decay of ^{241}Pu. Note that MOX containing ^{236}Pu at concentrations greater than about 10^{-4} weight percent of total plutonium mass or significant ^{241}Am in-growth may have larger gamma emission rates than are shown in Table C-4b.

Table C-4b.   Ratio of Gamma Emission Rate for Gamma Energies Exceeding 100 keV for MOX-Fresh Fuel Composed of Various Grades of Plutonium Relative to LEU-Fresh Fuel

Decay Time	No Plutonium	Pure ^{239}Pu	Weapon Grade	Fuel Grade	Power Grade
Initial	1.0	6.1	6.1	6.9	15.4
Maximum	1.0	6.1	6.1	7.2	83.5

The gamma emission rates for MOX-fresh fuel from both WG and FG plutonium are less than an order of magnitude greater than those for LEU-fresh fuel. The gamma emission rates for MOX-fresh fuel from PG plutonium can be up to about two orders of magnitude greater than those for LEU-fresh fuel depending on the time since MOX-fuel fabrication.

The neutron emission code SOURCES[C-4] can be used to determine the neutron emission rates for spontaneous fission and alpha-induced neutrons for equal weights of various nuclides of uranium, plutonium, and americium. Table C-5a lists the neutron emission rates for spontaneous fission and alpha-induced neutrons from ^{17}O and ^{18}O, for equal weights of nuclides at the initial MOX fuel fabrication time relative to the LEU neutron emission rate. Also listed in the table is the average neutron energy for each nuclide and each neutron emission process.

Table C-5a.   Neutron Emission Rates Relative to the LEU Neutron Emission Rate and Average Gamma Energies for Equal Weights of Some Nuclides of Uranium, Plutonium and Americium for (α,n) with ^{17}O and ^{18}O, Spontaneous Fission (SF), and the Sum of All Three (Total) Neutron Emission Processes

Select Nuclides	^{17}O (α,n) Relative to LEU	Average Energy (MeV)	^{18}O (α,n) Relative to LEU	Average Energy (MeV)	SF Relative to LEU	Average Energy (MeV)	Total Relative to LEU
^{238}Pu	1.03E+08	2.52	1.27E+08	2.37	1.98E+05	2.02	1.24E+06
^{239}Pu	2.86E+05	2.44	3.62E+05	2.25	1.67E+00	2.07	2.97E+03
^{240}Pu	1.06E+06	2.44	1.33E+06	2.25	7.84E+04	1.93	8.87E+04
^{241}Pu	9.42E+03	2.39	1.23E+04	2.19	3.77E+00	2.00	1.04E+02
^{242}Pu	1.49E+04	2.38	1.94E+04	2.19	1.31E+05	1.96	1.30E+05
^{233}U	3.50E+04	2.37	4.53E+04	2.17	6.23E-02	2.02	3.72E+02
^{235}U	5.96E+00	2.27	6.67E+00	2.07	2.29E-02	1.89	7.80E-02
^{236}U	1.87E+02	2.29	2.23E+02	2.09	4.19E-01	1.83	2.26E+00
^{238}U	7.93E-01	2.20	7.64E-01	1.97	1.04E+00	1.69	1.04E+00
LEU	1.00E+00	2.22	1.00E+00	2.00	1.00E+00	1.74	1.00E+00
^{241}Am	2.05E+07	2.51	2.53E+07	2.36	9.03E+01	2.15	2.09E+05

On an equal weight basis, ^{238}Pu, ^{240}Pu, ^{242}Pu and ^{241}Am are overwhelming the largest source for neutron emission for the nuclides listed in Table C-5a. Most nuclides listed in the table have neutron emission rates greater than LEU by one or more orders of magnitude. Table C-5a also shows that neutron emission from uranium isotopes is insignificant relative to that from plutonium isotopes on an equal weight basis. The average neutron energies listed in Table C-5a are between about 1.7 MeV and 2.5 MeV. This means that the spectral energy distribution for neutrons plays a much smaller role than does the spectral energy distribution for gammas.

The neutron emission code SOURCES can also be used to determine the neutron emission rates for spontaneous fission and alpha-induced neutrons for arbitrary compositions of plutonium isotopes with depleted uranium in MOX-fresh fuel. Table C-5b lists the ratio of neutron emission rates for MOX-fresh fuel relative to LEU-fresh fuel using the composition weight percents for MOX-fresh fuel fabricated from the various plutonium grades from Table C-2 in SOURCES. Note that MOX-fresh fuel with significant ^{241}Am in-growth (from ^{241}Pu decay) can have significantly larger relative neutron emission rates as is shown in the last line of Table C-5b.

Table C-5b.   Ratio of Neutron Emission Rate for MOX-Fresh Fuel Composed of Various Grades of Plutonium Relative to LEU-Fresh Fuel

Nuclide Composition	No Plutonium	Pure ^{239}Pu	Weapons Grade	Fuel Grade	Power Grade
Fresh Fuel	1	24	243	506	1,686
^{241}**Pu replaced by** ^{241}**Am**	1	24	250	536	1,995

Replacing 4 wt% ^{235}U with 4 wt% ^{239}Pu increases the neutron emission rate by a factor of about 24. Using WG plutonium instead of pure ^{239}Pu causes the neutron emission rate to increase by about another order of magnitude. Using FG plutonium instead of WG plutonium causes the neutron emission rate to increase by about another factor of two. Using PG plutonium instead of FG plutonium causes the neutron emission rate to increase by about another factor of three.

Plutonium-241 decays to ^{241}Am with a half-life of 14.35 years. Americium-241 is a stronger neutron source, so to get a bounding value for the expected increase in neutron emission rate, when ^{241}Pu decays to ^{241}Am, all ^{241}Pu is replaced with ^{241}Am, and the neutron emission rate is recalculated for each of these new artificial grades of plutonium.[*] The last row of Table C-5b lists the values obtained. This approach gives an indication of what decay time can do to neutron emission rates. The effect on neutron emission rate of ^{241}Pu decay to ^{241}Am is expected to be rather small except for MOX-fresh fuel fabricated from PG plutonium, where it could increase by a factor of about 20%.

The uncertainties in the rates of heat generation, or gamma emission, or neutron emission from analyses performed using radiation transport codes and cross section sets, such as those employed above, for MOX-RAM packages should be comparable to those performed for packages containing LEU RAM for the purposes required for thermal and shielding reviews.

In summary, heat generation and neutron emission rates increase significantly when MOX RAM replaces LEU RAM. The alpha-energy spectrum responsible for most heat generation is somewhat different for MOX RAM and LEU RAM, but that is not significant in relation to the magnitude of the heat generation rates between them. The neutron energy spectrum from MOX RAM and LEU RAM are also somewhat different, but again this is not significant in relation to the magnitude of the neutron emission rates between them. The gamma emission rate increases between MOX RAM and LEU RAM are not as important as long as ^{236}Pu is less than about 10^{-4} weight percent of the heavy metal present in MOX RAM. Otherwise, the strong 2.614 MeV gamma from the chain decay of ^{236}Pu to ^{208}Tl becomes an important source of external package gamma dose that is extremely hard to shield against. However, the gamma energy spectrum from MOX RAM and LEU RAM can be quite different depending on the nuclides present and this can be significant from a shielding point of view.

---

[*]   Replacing ^{241}Pu with ^{241}Am is bounding for a neutron shielding evaluation, but not for a criticality evaluation.

References

C-1. U.S. Department of Energy, "Stabilization, Packaging, and Storage of Plutonium-Bearing Materials," U.S. DOE Standard DOE-STD-3013-2000, Washington D.C., September 2000.

C-2. S. M. Bowman and L. C. Leal, "ORIGEN-ARP: Automatic Rapid Process for Spent Fuel Depletion, Decay, and Source Term Analysis," part of *SCALE: A Modular Code System for Performing Standardized Computer Analyses for Licensing Evaluation*, Vol. I, NUREG/CR-0200, Revision 6, (ORNL/NUREG/CSD-2/VI/R6), Oak Ridge National Laboratory, March 2000.

C-3. T. B. Gosnell, "Automated Calculation of Photon Source Emission From Arbitrary Mixtures of Naturally Radioactive Nuclides," *Nuclear Instruments and Methods in Physics Research*, **A299**, 682–686 (1990).

C-4. W. B. Wilson, et al., "SOURCES 4A: A Code for Calculating (α,n), Spontaneous Fission, and Delayed Neutron Sources and Spectra," LA-13639-MS, Los Alamos National Laboratory, Los Alamos, NM, September 1999.

APPENDIX D: BENCHMARK CONSIDERATIONS FOR MOX RADIOACTIVE MATERIALS

D.1 Experimental Benchmarks

Substantial guidance on how to select an appropriate set of criticality benchmark experiments for LEU fissile systems is given in NUREG/CR-5661 and in NUREG/CR-6361.[D-1, D-2] Considerably fewer benchmark experiments exist for MOX than for LEU, however. As a consequence, the guidance provided in NUREG/CR-5661 and/or in NUREG/CR-6361 cannot be applied directly to the evaluation of MOX fissile systems. The benchmarks needed for the criticality analyses of MOX packages are in the thermal energy range. This condition results because, for essentially all types of MOX, the most reactive configuration is a flooded containment.

As an alternative, the International Handbook of Evaluated Criticality Safety Benchmark Experiments[D-3] (IHECSBE) has 11 evaluated thermal-energy studies involving MOX fuel pins in various lattice experiments, and 5 evaluated thermal-energy studies involving MOX liquids in tank experiments. These can be divided into 18 sets of experiments involving different fissile oxide compositions and configurations in lattices, and 13 sets of experiments involving different liquid fissile nitrate compositions and configurations in tanks. The total number of essentially different experiments is 131. Other benchmark experiments are available throughout the world, but are not as readily available, and the vast majority have not been rigorously evaluated in the manner of those found in the IHECSBE, and are consequently of limited use for benchmark criticality analyses for MOX packages. More evaluated MOX thermal benchmarks are expected in future editions of the IHECSBE.

The 18 sets of experiments involving fissile oxides in lattices and 13 sets of experiments involving fissile nitrate liquids in tanks have been organized and shown in Tables D-1 through D-5. The various tables are separated on two features. The first is between lattice and tank experiments, and the second is on weight percent of plutonium to total plutonium plus uranium, Pu/(Pu+U). Table D-1 has lattice experiments with Pu/(Pu+U) to 5%. Table D-2 has lattice experiments with Pu/(Pu+U) from 5% to 15%. Table D-3 has lattice experiments with Pu/(Pu+U) greater than 15%. Table D-4 has tank experiments with Pu/(Pu+U) to 31% (there are no experiments with Pu/(Pu+U) less than 22%). Table D-5 has tank experiments with Pu/(Pu+U) greater than 31%. Lists of meaningful, experimental characteristics are recorded for each set of experiments together with characteristics of their corresponding computational evaluations.

Experimental plutonium benchmarks should also be taken into account as part of the initial set of benchmark experiments to be considered for a MOX package application. About four times as many thermal-plutonium-tank-liquid benchmarks exist in the IHECSBE as thermal-MOX-tank-liquid benchmarks. However, fewer thermal-plutonium-lattice benchmarks exist in the IHECSBE as thermal-MOX-lattice benchmarks.

D.2 Summary of Bias and Uncertainty Evaluation

There are two measures to determine the accuracy of an experiment and its associated calculation. The first measure is the effective bias (Eff-Bias) between calculation and benchmark experiment. The multiplication coefficient for a fissile system is designated as k_{eff}. Designate the calculated k_{eff} for the benchmark experiment as k_{calc} and the benchmark experimental k_{eff} as k_{exp}. If the calculational bias, β, is defined as $\beta = k_{calc} - k_{exp}$, then a quantity Δk^* can be defined as

---

*   As defined in Equation D-1, Δk is always less than or equal to zero, and is consistent with the bias, $\bar{\beta}$, defined in Reference D-1. Typically, a calculational method is termed to have a negative bias if it under-predicts the critical condition.

$$\Delta k = \begin{pmatrix} \beta & \text{if } k_{calc} \le k_{exp} \\ \\ 0 & \text{if } k_{calc} > k_{exp} \end{pmatrix} . \qquad (D\text{-}1)$$

For a given experimental benchmark set, Δk_{max} is chosen as the largest absolute value of the Δk given by Equation D-1 for all experiments in the set. The 95% confidence limit of k_{calc} is k_{calc} plus twice the calculated standard deviation, which is designated by 2σ. The Eff-Bias value is then given by

$$\text{Eff-Bias} = \Delta k_{max} - 2\sigma. \qquad (D\text{-}2)$$

Eff-Bias, as defined here, is always *less* than zero. If k_{calc} is greater than k_{exp} for all experiments in a set, the Eff-Bias value is just the negative of twice the calculated standard deviation.

The second measure is the total experimental uncertainty (Exp-Uncer) that was determined by the evaluator after assessing all sources of uncertainty for the experiments in a set.[*] A worst-case difference between k_{calc} and k_{exp} can be assigned as the difference of the total experimental uncertainty and the effective bias (Exp-Uncer – Eff-Bias) for the experimental set in question. This worst-case difference (WCD), as defined here, is always *greater* than zero. It represents the upper limit of the inherent uncertainties in the ability of the computer code, together with the cross-section set used, to accurately determine the k_{eff} of a critical benchmark experiment. Therefore, a bounding multiplication coefficient, k_{safe}, at the 95% confidence limit, can be chosen to be equal to 0.95 minus WCD, where an administrative margin of safety of 0.05 has been included.[+]

Values for the variable WCD for each experimental set vary between 0.0071 to 0.0192 (0.71% to 1.92%), 0.0043 to 0.0328 (0.43% to 3.28%), 0.0023 to 0.0138 (0.23% to 1.38%), 0.0044 to 0.0180 (0.44% to 1.80%), and 0.0044 to 0.0150 (0.44% to 1.50%), for the experimental sets in Tables D-1, D-2, D-3, D-4, and D-5, respectively. No particular correlation seems to exist between WCD and the lattice configuration or pitch. Neither does there seem to be a correlation with plutonium composition type (Pu type). The plutonium composition types are given in Table 1 in the text, and are designated as weapons grade (WG), fuel grade (FG), and power grade (PG).

The maximum value for WCD found in the five tables is 0.0328 or 3.28% in k_{eff}. How accurately a criticality computer code can predict the critical value for a criticality experiment depends on the methodology employed by the code and the cross-section set used, together with the detail to which the experimental system is modeled in the computer. In addition, the basic experimental uncertainty limits the ultimate prediction accuracy possible. Of particular importance is the cross-section set. Values for WCD in the five tables that are significantly less than 0.0100 are due to the fact that k_{calc} is greater than k_{exp}. Therefore, the value for Eff-Bias, in that case, is just the negative of twice the calculated standard deviation, which is approximately 0.0020. Cross-section sets prior to ENDF/B-V over predict plutonium reactivity, and this represents some of the reason for the over-prediction for k_{calc} for these experiments. Values for k_{safe} are not expected to be much above 0.93, except when it can be demonstrated that the criticality code and cross-section set overestimates the reactivity of the MOX contents.

Analyzing an acceptable number of MOX benchmarks is the preferred way to obtain a bias value for the MOX contents of a package. With the relatively limited number of MOX-critical experiments available for use in validation exercises, it is important to determine that the application of interest to the reviewer fits within the area of applicability for the set of critical benchmark experiments selected for validation. Guidance on how to select an appropriate set of benchmark experiments for LEU fissile system is given in NUREG/CR-5661 and in

---

[*]   The evaluator included sources of experimental bias or error in each k_{exp}. This does not represent an uncertainty and so is not included in the value for total experimental uncertainty.

[+]   If the benchmarks are applied to a package application where there is a lack of experimental data, the 0.05 administrative margin may not be sufficient, and the reviewer needs to be aware of this issue. In reality, the 0.05 margin should be sufficient, but there needs to be an assessment of the adequacy of the 0.05 to establish the basis. Guidance for deciding on an acceptable choice for the administrative margin is given in NUREG/CR-5561. See also NUREG/CR-6361.

NUREG/CR-6361. An important advancement using computational methodology to select an appropriate set of benchmark experiments for a fissile package application is currently being developed for SCALE.[D-4, D-5, D-6]

A set of sensitivity and uncertainty analysis tools are being developed for version 5 of SCALE that gives a measure of the similarity of the reactivity of a package application to that of an experimental benchmark. Sensitivity coefficients for both systems are computed and give the sensitivity of each system's k_{eff} to the cross section data. These sensitivity coefficients are determined for each energy group in the cross section library chosen in the analysis, as well as the sum over all energy groups. Two integral parameters for the combined systems are produced from the sensitivity data to determine system-to-system similarities. The first parameter can be used as a gauge of system similarity to sensitivity-only. The second parameter can be used as a measure of the similarity of the systems in terms of uncertainty, not just sensitivity. The pair of integral parameter values is determined for every potential benchmark experiment with the package application of interest. When two systems produce a value of 0.8 for either integral parameter, or both, this indicates the k_{eff} response is similar enough that one system serves well to validate the criticality safety parameters for the other system. The benchmark experiments chosen for complete validation are those with high integral parameter values.[D-4, D-5, D-6]

New parameters can also be constructed from the components of the integral parameters and can be used to explore the sensitivity of specific nuclide reactions of benchmark experiments with the package application of interest. For example, if low integral parameter values are found for an application with all benchmark experiments chosen for validation, the new parameters could serve to identify which nuclides would require additional experimental benchmark data for complete validation. Also, in the validation of shipping casks for commercial fuel, numerous benchmark experiments might serve to validate the fission reactions, and thus high integral parameter values would be found. However, the new parameters could be used to find benchmarks to ensure that any poison materials in the cask are also well validated by the benchmarks. Once these sensitivity and uncertainty analysis tools are released with version 5 of SCALE, the criticality safety analyst will have a powerful set of tools to perform detailed quantitative analyses to determine the applicability of benchmark experiments to help design package applications under consideration.[D-4, D-5, D-6]

Table D-1. Important Characteristics of Lattice Experiments with Weight Percent of Pu/(Pu+U) to 5% (from IHECSBE)

	MCT-009	MCT-002	MCT-002	MCT-006	MCT-007	MCT-008	MCT-004	MCT-005
Designation for experiments[a]	MCT-009	MCT-002	MCT-002	MCT-006	MCT-007	MCT-008	MCT-004	MCT-005
Facility where experiments conducted	Hanford	Hanford	Hanford	Hanford	Hanford	Hanford	Tokai	Hanford
Computer codes used in evaluations[b]	MCNP/KENO	MCNP	MCNP	MCNP/KENO	MCNP/KENO	MCNP/KENO	MCNP/KENO	MCNP/KENO
Cross-section sets used in evaluations[c]	ENDF/B-V/IV	ENDF/B-V	ENDF/B-V	ENDF/B-V/IV	ENDF/B-V/IV	ENDF/B-V/IV	JENDL-3.2	ENDF/B-IV&V
Cross-section type[d]	cont/27grp	cont	cont	cont/27grp	cont/27grp	cont/27grp	cont/137grp	cont/27grp
Fuel compound[e]	oxide	oxide	oxide	oxide	oxide	oxide	oxide	oxide
Fuel compound form	solid	solid	solid	solid	solid	solid	solid	solid
Density of fuel[f]	86.7%	86.7%	86.7%	86.7%	86.7%	86.7%	55%	86%
Organization of fuel[g]	pins	pins	pins	pins	pins	pins	pins	pins
Cladding used for fuel[h]	Zirc-2	Zirc-2	Zirc-2	Zirc-2	Zirc-2	Zirc-2	Zirc-2	Zirc-2
Pu/(Pu+U) atom percent	1.51%	1.80%	1.80%	1.80%	2.01%	2.01%	3.03%	3.52%
U-235 atom percent	0.16%	0.71%	0.71%	0.71%	0.72%	0.72%	0.71%	0.71%
U-238 atom percent	99.84%	99.29%	99.29%	99.29%	99.28%	99.28%	99.29%	99.29%
Pu-238 atom percent	-	0.01%	0.01%	0.01%	-	-	0.50%	0.28%
Pu-239 atom percent	91.41%	91.84%	91.84%	91.84%	81.11%	71.76%	68.18%	75.39%
Pu-240 atom percent	7.83%	7.76%	7.76%	7.76%	16.54%	23.50%	22.02%	18.10%
Pu-241 atom percent	0.73%	0.37%	0.37%	0.37%	2.15%	4.08%	7.26%	5.08%
Pu-242 atom percent	0.03%	0.03%	0.03%	0.03%	0.20%	0.66%	2.04%	1.15%
Plutonium type as given in Table 1	WG	WG	WG	WG	FG	PG	PG	FG-PG
Shape of lattice[i]	cylinder	rectangle	rectangle	cylinder	cylinder	cylinder	rectangle	cylinder
Pitch of lattice	triangle	square	square	triangle	triangle	triangle	square	triangle
Number of experiments in each set	6	3	3	6	5	6	4	7
Fissile moderator used[j]	H₂O	H₂O	B-H₂O	H₂O	H₂O	H₂O	H₂O	H₂O
Reflector used	H₂O	H₂O	B-H₂O	H₂O	H₂O	H₂O	H₂O	H₂O
Maximum effective bias of experiments in set (Eff-Bias)	-0.0112	-0.0052	-0.0026	-0.0089	-0.0040	-0.0068	-0.0097	-0.0037
Maximum uncertainty of experiments in set (Exp-Uncer)	0.0080	0.0059	0.0045	0.0054	0.0061	0.0065	0.0051	0.0042
Exp-Uncer minus Eff-Bias (WCD)	0.0192	0.0111	0.0071	0.0143	0.0101	0.0133	0.0148	0.0079

a Definition of acronyms is MCT = MIX-COMP-THERM
b Codes MCNP[b.7] and KENO[b.8]
c ENDF/B-V/IV means cross-section set ENDF/B-V for MCNP and cross-section set ENDF/B-IV for KENO J ENDL-3 2 is cross-section set for both MCNP and KENO
d Cross-section type is either continuous cross sections (cont) or group cross sections (27grp, 137grp)
e Heavy metal is an oxide
f MOX density given as percent of theoretical density taken as 11 00 g/cm³
g Pins means organization of MOX as pellets in fuel pins
h Zirc-2 means zircaloy-2 cladding
i Cylinder means shape of lattice is a cylinder Rectangle means shape of lattice is a rectangle
j B-H₂O means borated water as moderator or reflector

D-4

Table D-2. Important Characteristics of Lattice Experiments with Weight Percent of Pu/(Pu+U) from 5% to 15% (from IHECSBE)

Designation for experiments[a]	MCT-012	MCT-012	MCT-012	MCT-012	MCT-003	MCT-003
Facility where experiments conducted	Hanford	Hanford	Hanford	Hanford	WREC	WREC
Computer codes used in evaluations[b]	MCNP/KENO	MCNP/KENO	MCNP/KENO	MCNP/KENO	MCNP	MCNP
Cross-section sets used in evaluations[c]	ENDF/B-V	ENDF/B-V	ENDF/B-V	ENDF/B-V	ENDF/B-V	ENDF/B-V
Cross-section type[d]	cont/238grp	cont/238grp	cont/238grp	cont/238grp	cont	cont
Fuel compound[e]	oxide-poly	oxide-poly	oxide-poly	oxide-poly	oxide	oxide
Fuel compound form	solid	solid	solid	solid	solid	solid
Density of fuel[f]	N/A	N/A	N/A	N/A	94%	94%
Organization of fuel[g]	cubes, slabs	cubes, slabs	cubes, slabs	cubes, slabs	pins	pins
Cladding used for fuel[h]	plastic 471	plastic 471	plastic 471	plastic 471	Zirc-4	Zirc-4
Pu/(Pu+U) atom percent	14.62%	14.62%	7.89%	7.60%	6.63%	6.63%
U-235 atom percent	0.15%	0.15%	0.15%	0.15%	0.71%	0.71%
U-238 atom percent	99.85%	99.85%	99.85%	99.85%	99.29%	99.29%
Pu-238 atom percent	-	-	-	0.59%	-	-
Pu-239 atom percent	91.42%	91.42%	91.25%	67.97%	90.65%	90.65%
Pu-240 atom percent	7.97%	7.97%	8.12%	22.95%	8.55%	8.55%
Pu-241 atom percent	0.57%	0.57%	0.58%	5.57%	0.76%	0.76%
Pu-242 atom percent	0.04%	0.04%	0.05%	2.92%	0.04%	0.04%
Plutonium type as given in Table 1	WG	WG	WG	PG	WG-FG	WG-FG
Shape of lattice[i]	3D cube	3D cube	3D cube	3D cube	rectangle	rectangle
Pitch of lattice	square	square	square	square	square	square
Number of experiments in each set	3	6	7	6	1	5
Fissile moderator used[j]	polystyrene	polystyrene	polystyrene	polystyrene	B-H$_2$O	H$_2$O
Reflector used	none	plexiglas	plexiglas	plexiglas	B-H$_2$O	H$_2$O
Maximum effective bias of experiments in set (Eff-Bias)	-0.0020	-0.0016	-0.0016	-0.0270	-0.0030	-0.0063
Maximum uncertainty of experiments in set (Exp-Uncer)	0.0037	0.0027	0.0036	0.0058	0.0052	0.0071
Exp-Uncer minus Eff-Bias (WCD)	0.0057	0.0043	0.0052	0.0328	0.0082	0.0134

a Definition of acronyms is MCT = MIX-COMP-THERM
b Codes MCNP[D-7] and KENO[D-8]
c ENDF/B-V is cross-section set for MCNP and KENO
d Cross-section type is either continuous cross sections (cont) or group cross sections (238grp)
e Heavy metal is an oxide   Oxide-poly means mixture of MOX particles and polystyrene pressed into cubes and slabs
f MOX density given as percent of theoretical density taken as 11 00 g/cm^3
g Pins means organization of MOX as pellets in fuel pins   Cubes, slabs means organization of MOX-polystyrene is as cubes and slabs
h Zirc-4 means zircaloy-4 cladding   Plastic 471 means cladding is six mil plastic tape MM&M (3M) #471
i Rectangle means shape of lattice is a rectangle   3D cube means cubes and slabs stacked into the shape of a 3D-rectangular cube
j B-H$_2$O means borated water as moderator or reflector

D-5

Table D-3. Important Characteristics of Lattice Experiments with Weight Percent of Pu/(Pu+U) Greater than 15% (from IHECSBE)

Designation for experiments[a]	MCT-001	MCT-011	MCT-012	MCT-012	MCT-012
Facility where experiments conducted	Hanford	Valduc	Hanford	Hanford	Hanford
Computer codes used in evaluations[b]	MONK	MORET	MCNP/KENO	MCNP/KENO	MCNP/KENO
Cross-section sets used in evaluations[c]	UKNDL	JEF2.2	ENDF/B-V	ENDF/B-V	ENDF/B-V
Cross-section type[d]	cont	172gp	cont/238grp	cont/238grp	cont/238grp
Fuel compound[e]	oxide	oxide	oxide-poly	oxide-poly	oxide-poly
Fuel compound form	solid	solid	solid	solid	solid
Density of fuel[f]	89.4%	94.2%	N/A	N/A	N/A
Organization of fuel[g]	pins	pins	cubes, slabs	cubes, slabs	cubes, slabs
Cladding used for fuel[h]	316 SS	Z3CND18.12 SS	plastic 471	plastic 471	plastic 471
Pu/(Pu+U) atom percent	19.70%	25.80%	30.00%	30.00%	30.00%
U-235 atom percent	0.71%	60.15%	0.15%	0.15%	0.15%
U-238 atom percent	99.29%	39.85%	99.85%	99.85%	99.85%
Pu-238 atom percent	0.15%	-	-	-	-
Pu-239 atom percent	85.54%	89.00%	91.22%	91.22%	91.22%
Pu-240 atom percent	11.46%	9.72%	8.13%	8.13%	8.13%
Pu-241 atom percent	2.50%	1.21%	0.61%	0.61%	0.61%
Pu-242 atom percent	0.35%	0.07%	0.04%	0.04%	0.04%
Plutonium type as given in Table 1	FG	WG-FG	WG	WG	WG
Shape of lattice[i]	rectangle	cylinder	3D cube	3D cube	3D cube
Pitch of lattice	square	triangle	square	square	square
Number of experiments in each set	4	6	8		3
Fissile moderator used	H_2O	H_2O	polystyrene		polystyrene
Reflector used	H_2O	H_2O	plexiglas		none
Maximum effective bias of experiments in set (Eff-Bias)	-0.0103	-0.0006	-0.0018		-0.0086
Maximum uncertainty of experiments in set (Exp-Uncer)	0.0025	0.0017	0.0049		0.0052
Exp-Uncer minus Eff-Bias (WCD)	0.0128	0.0023	0.0067		0.0138

a. Definition of acronyms is MCT = MIX-COMP-THERM.
b. Codes MCNP[D-7] and KENO[D-8] MONK,[D-9] and MORET[D-10]
c. ENDF/B-V is cross-section set for MCNP and KENO. UKNDL is cross-section set for MONK. JEF2.2 is cross-section set for MORET.
d. Cross-section type is either continuous cross sections (cont.) or group cross sections (172grp, 238grp)
e. Heavy metal is an oxide. Oxide-poly means mixture of MOX particles and polystyrene pressed into cubes and slabs.
f. MOX density given as percent of theoretical density taken as 11.00 g/cm³.
g. Pins means organization of MOX is as pellets in fuel pins. Cubes, slabs means organization of MOX-polystyrene is as cubes and slabs.
h. SS means stainless steel cladding. Plastic 471 means cladding is six mil plastic tape MM&M (3M) #471.
i. Cylinder means shape of lattice is a cylinder. Rectangle means shape of lattice is a rectangle cube. 3D cube means cubes and slabs stacked into the shape of a 3D-rectangular cube.

Table D-4. Important Characteristics of Tank Experiments with Weight Percent of Pu/(Pu+U) to 31% (from IHECSBE)

Designation for experiments[a]	MST-001	MST-001	MST-001	MST-001	MST-002	MST-003
Facility where experiments conducted	Hanford	Hanford	Hanford	Hanford	Hanford	AWRE
Computer codes used in evaluations[b]	MCNP/KENO	MCNP/KENO	MCNP/KENO	MCNP/KENO	MCNP/KENO	MONK
Cross-section sets used in evaluations[c]	ENDF/B-V /IV	ENDF/B-V /IV	ENDF/B-V /IV	ENDF/B-V /IV	ENDF/B-V /IV	UKNDL
Cross-section type[d]	cont/27grp	cont/27grp	cont/27grp	cont/27grp	cont/27grp	cont/27grp
Fuel compound[e]	nitrate	nitrate	nitrate	nitrate	nitrate	nitrate
Fuel compound form	liquid	liquid	liquid	liquid	liquid	liquid
Density of fuel[f]	1.31-1.68	1.31-1.68	1.31-1.48	1.70	1.09	1.11-1.52
Pu/(Pu+U) atom percent	22%	22%	22%	22%	23%	30.7%
U-235 atom percent	0.70%	0.70%	0.70%	0.70%	0.70%	0.72%
U-238 atom percent	99.30%	99.30%	99.30%	99.30%	99.30%	99.28%
Pu-238 atom percent	0.03%	0.03%	0.03%	0.03%	0.03%	-
Pu-239 atom percent	91.12%	91.12%	91.12%	91.12%	91.12%	93.95%
Pu-240 atom percent	8.34%	8.34%	8.34%	8.34%	8.31%	5.63%
Pu-241 atom percent	0.42%	0.42%	0.42%	0.42%	0.45%	0.42%
Pu-242 atom percent	0.09%	0.09%	0.09%	0.09%	0.09%	-
Plutonium type as given in Table 1	WG	WG	WG	WG	WG	WG
Tank fissile liquid is in[g]	N/A	cylinder	cylinder	cylinder	cylinder	slab
Auxiliary tank additional fissile liquid is in[h]	annular	annular	annular	N/A	N/A	N/A
Number of experiments in each set	2	5	2	1	1	10
Fissile moderator used[i]	soln H$_2$O	soln H$_2$O	soln H$_2$O	soln H$_2$O	soln H$_2$O	soln H$_2$O
Reflector used[j]	B$_4$C-concrete	B$_4$C-concrete	poly-Cd cover	none	H$_2$O	H$_2$O & poly
Maximum effective bias of experiments in set (Eff-Bias)	-0.0101	-0.0164	-0.0028	-0.0068	-0.0020	-0.0038
Maximum uncertainty of experiments in set (Exp-Uncer)	0.0016	0.0016	0.0016	0.0016	0.0024	0.0025
Exp-Uncer minus Eff-Bias (WCD)	0.0117	0.0180	0.0044	0.0084	0.0044	0.0063

a. Definition of acronyms is MST = MIX-SOL-THERM.
b. Codes MCNP,[D-7] KENO,[D-8] and MONK.[D-9]
c. ENDF/B-V/IV means cross-section set ENDF/B-V for MCNP and ENDF/B-IV for KENO. UKNDL is cross-section set for MONK.
d. Cross-section type is either continuous cross sections (cont.) or group cross sections (27grp).
e. Heavy metal is as a nitrate dissolved in dilute nitric acid solution.
f. Solution density is in g/ml.
g. Containers for fissile solution are cylinders or slabs.
h. Annular tank surrounding central cylindrical tank or just an annular tank.
i. Soln H$_2$O means the moderator is the fissile nitrate solution.
j B$_4$C-concrete means borated concrete. Poly-Cd cover means polyethylene reflector coated with Cd.

Table D-5. Important Characteristics of Tank Experiments with Weight Percent of Pu/(Pu+U) Greater than 31% (from IHECSBE)

Designation for experiments[a]	MST-004	MST-004	MST-004	MST-005	MST-005	MST-002	MST-001
Facility where experiments conducted	Hanford	Hanford	Hanford	Hanford	Hanford	Hanford	Hanford
Computer codes used in evaluations[b]	MCNP/KENO	MCNP/KENO	MCNP/KENO	MCNP/KENO	MCNP/KENO	MCNP/KENO	MCNP/KENO
Cross-section sets used in evaluations[c]	ENDF/B-V /IV	ENDF/B-V /IV	ENDF/B-V /IV	ENDF/B-V /IV	ENDF/B-V /IV	ENDF/B-V /IV	ENDF/B-V /IV
Cross-section type[d]	cont/27grp	cont/27grp	cont/27grp	cont/27grp	cont/27grp	cont/27grp	cont/27grp
Fuel compound[e]	nitrate	nitrate	nitrate	nitrate	nitrate	nitrate	nitrate
Fuel compound form	liquid	liquid	liquid	liquid	liquid	liquid	liquid
Density of fuel[f]	1.17-1.67	1.17-1.67	1.17-1.67	1.17-1.67	1.17-1.67	1.05	1.15-1.44
Pu/(Pu+U) atom percent	40%	40%	40%	40%	40%	52%	97%
U-235 atom percent	0.56%	0.56%	0.56%	0.56%	0.56%	0.70%	2.29%
U-238 atom percent	99.44%	99.44%	99.44%	99.44%	99.44%	99.30%	97.71%
Pu-238 atom percent	0.03%	0.03%	0.03%	0.03%	0.03%	0.03%	0.03%
Pu-239 atom percent	91.12%	91.12%	91 12%	91.12%	91.12%	91.12%	91.57%
Pu-240 atom percent	8.34%	8.34%	8.34%	8.34%	8.34%	8.34%	7.94%
Pu-241 atom percent	0.42%	0.42%	0.42%	0.42%	0.42%	0.42%	0.39%
Pu-242 atom percent	0.09%	0.09%	0.09%	0.09%	0.09%	0.09%	0.07%
Plutonium type as given in Table 1	WG	WG	WG	WG	WG	WG	WG
Tank fissile liquid is in[g]	cylinder	cylinder	cylinder	slab	slab	cylinder	cylinder
Auxiliary tank additional fissile liquid is in[h]	N/A	N/A	N/A	N/A	N/A	N/A	annular
Number of experiments in each set	3	3	3	3	4	2	3
Fissile moderator used[i]	soln H_2O	soln H_2O	soln H_2O	soln H_2O	soln H_2O	soln H_2O	soln H_2O
Reflector used[j]	none	H_2O	concrete	none	H_2O	H_2O	B_4C-concrete
Maximum effective bias of experiments in set (Eff-Bias)	-0.0060	-0.0048	-0.0024	-0.0114	-0.0026	-0.0020	-0.0032
Maximum uncertainty of experiments in set (Exp-Uncer)	0.0033	0.0033	0.0078	0.0036	0.0037	0.0024	0.0016
Exp-Uncer minus Eff-Bias (WCD)	0.0093	0.0081	0.0102	0.0150	0.0063	0.0044	0.0048

a. Definition of acronyms is MST = MIX-SOL-THERM.

b. Codes MCNP[D-7] and KENO.[D-8]

c. ENDF/B-V/IV means ENDF/B-V for MCNP and ENDF/B-IV for KENO.

d. Cross-section type is either continuous cross sections (cont.) or group cross sections (27grp).

e. Heavy metal is as a nitrate dissolved in dilute nitric acid solution.

f. Solution density is in g/ml.

g. Containers for fissile solution are cylinders or slabs.

h. Annular tank surrounding central cylindrical tank.

i. Soln H_2O means the moderator is the fissile nitrate solution.

j. B_4C-concrete means borated concrete.

References

D-1. H. R. Dyer and C. V. Parks, "Recommendations for Preparing the Criticality Safety Evaluation of Transportation Packages," NUREG/CR-5661, April 1997.

D-2 J. J. Lichtenwalter, S. M. Bowman, M. D. DeHart, and C. M. Hopper, "Criticality Benchmark Guide for Light-Water-Reactor Fuel in Transportation and Storage Packages," NUREG/CR-6361, March 1997.

D-3. Nuclear Energy Agency, "International Handbook of Evaluated Criticality Safety Benchmark Experiments," Organization for Economic Co-operation and Development, NEA/NSC/DOC(95)03, September 2001 Edition.

D-4. B. L. Broadhead, C. M. Hopper, R. L. Childs, and C. V. Parks, "Sensitivity and Uncertainty Analyses Applied to Criticality Safety Validation," NUREG/CR-6655, Vols. 1 and 2 (ORNL/TM-13692/V1 and V2), U.S. Nuclear Regulatory Commission, Oak Ridge National Laboratory, November 1999.

D-5. B. T. Rearden and R. L. Childs, "Prototypical Sensitivity and Uncertainty Analysis Codes for Criticality Safety with the SCALE Code System," *Trans. Am. Nucl. Soc.*, Washington, D.C., November 2000.

D-6. M. E. Dunn and B. T. Rearden, "Application of Sensitivity and Uncertainty Analysis Methods to a Validation Study for Weapons-Grade Mixed-Oxide Fuel," 2001 ANS Embedded Topical Meeting on Practical Implementation of Nuclear Criticality Safety, Reno, NV, November 11–15, 2001.

D-7. MCNP is a three-dimensional, Monte Carlo radiation transport code that uses point-wise cross sections, developed by LANL. *MCNP – A General Monte Carlo N-Particle Transport Code*, Version 4B, Judith F. Briesmeister, Editor, Los Alamos Report, LA-12625-M, March 1997. Various versions of MCNP are available. This reference is for version 4B.

D-8. KENO is a three-dimensional Monte Carlo criticality module in the SCALE system that uses multigroup cross sections, developed by ORNL. *SCALE: A Modular Code System for Performing Standardized Analyses for Licensing Evaluations*, NUREG/CR-0200, Rev. 4 (ORNL/NUREG/CSD-2/R4), Vols. I, II, III, October 1995. Various versions of SCALE are available. This reference is for version 4.3.

D-9. MONK is a three-dimensional Monte Carlo radiation transport code that uses point-wise cross sections, developed by A.E.A Technology of the United Kingdom.

D-10. MORET is a three-dimensional Monte Carlo criticality code that uses multigroup cross sections, developed by C.E.A. of France.